T0328279

Responsible Genomic Data Sharing

Responsible Genomic Data Sharing

Challenges and Approaches

Edited by

Xiaoqian Jiang

Associate Professor
School of Biomedical Informatics
Houston, TX, United States

Haixu Tang

Professor
Department of Computer Science
Indiana University
Bloomington, IN, United States

Academic Press is an imprint of Elsevier
125 London Wall, London EC2Y 5AS, United Kingdom
525 B Street, Suite 1650, San Diego, CA 92101, United States
50 Hampshire Street, 5th Floor, Cambridge, MA 02139, United States
The Boulevard, Langford Lane, Kidlington, Oxford OX5 1GB, United Kingdom

Library of Congress Cataloging-in-Publication Data
A catalog record for this book is available from the Library of Congress

British Library Cataloguing-in-Publication Data
A catalogue record for this book is available from the British Library

ISBN: 978-0-12-816197-5

For information on all Academic Press publications visit our website at https://www.elsevier.com/books-and-journals

Publisher: Andre Gerhard Wolff
Acquisitions Editor: Peter B. Linsley
Editorial Project Manager: Samantha Allard
Production Project Manager: Sreejith Viswanathan
Cover Designer: Matthew Limbert

Typeset by TNQ Technologies

Contents

Contributors

Erman Ayday, PhD
Case Western Reserve University, Department of Electrical Engineering and Computer Science, Cleveland, OH, United States; Bilkent University, Ankara, Turkey

Somnath Chakrabarti
Intel Labs, Intel Corporation, Hillsboro, OR, United States

Ellen Wright Clayton
Center for Biomedical Ethics and Society, Vanderbilt University Medical Center, Nashville, TN, United States; Law School, Vanderbilt University, Nashville, TN, United States; Department of Pediatrics, Vanderbilt University Medical Center, Nashville, TN, United States

Stephanie O.M. Dyke, PhD
Academic Associate, McGill Centre for Integrative Neuroscience, Montreal Neurological Institute, Department of Neurology & Neurosurgery, Faculty of Medicine, McGill University, Montreal, QC, Canada

Gamze Gürsoy, PhD
Postdoctoral Research Associate, Program in Computational Biology and Bioinformatics, Department of Molecular Biophysics and Biochemistry, Yale University, New Haven, CT, United States

Yan Huang, PhD
Assistant Professor, Computer Science Indiana University, Bloomington, IN, United States

Murat Kantarcioglu
Department of Computer Science, University of Texas at Dallas, Richardson, TX, United States

Thomas Knauth
Intel Labs, Intel Corporation, Hillsboro, OR, United States

Dmitrii Kuvaiskii
Intel Labs, Intel Corporation, Hillsboro, OR, United States

Kim Laine, PhD
Microsoft Research, Redmond WA, United States

Ryan Lewis
School of Biomedical Informatics, University of Texas Health Science Center at Houston, Houston, TX, United States

Bradley Malin
Department of Electrical Engineering and Computer Science, Vanderbilt University, Nashville, TN, United States; Department of Biomedical Informatics, Vanderbilt University Medical Center, Nashville, TN, United States; Department of Biostatistics, Vanderbilt University Medical Center, Nashville, TN, United States

Ardalan Naseri, PhD
School of Biomedical Informatics, University of Texas Health Science Center at Houston, Houston, TX, United States

Michael Steiner
Intel Labs, Intel Corporation, Hillsboro, OR, United States

Mona Vij
Intel Labs, Intel Corporation, Hillsboro, OR, United States

Yevgeniy Vorobeychik
Department of Computer Science and Engineering, Washington University in St. Louis, St. Louis, MO, United States

Zhiyu Wan
Department of Electrical Engineering and Computer Science, Vanderbilt University, Nashville, TN, United States

Yuan Wei
Department of Computer Science, University of Central Florida, Orlando, FL, United States

Shaojie Zhang, PhD
Department of Computer Science, University of Central Florida, Orlando, FL, United States

Degui Zhi, PhD
School of Biomedical Informatics, University of Texas Health Science Center at Houston, Houston, TX, United States

SECTION I

Privacy challenges in genomic data sharing

CHAPTER

1

Criticality of data sharing in genomic research and public views of genomic data sharing

Gamze Gürsoy, PhD

Postdoctoral Research Associate, Program in Computational Biology and Bioinformatics, Department of Molecular Biophysics and Biochemistry, Yale University, New Haven, CT, United States

1. Introduction

DNA is the carrier of genetic information and is passed down from one generation to the next. With the completion of the Human Genome Project, this genetic information is more accessible now than ever before. The human genome that was released in June 2000 is a representative reference genome sequence based on the DNA taken from a small number of individuals [1]. Although Homo sapiens share a large portion of their genomes, as more individuals' genomes are studied [2,3], the variation between individual genomes is being identified. Variation in our genomes is what makes us unique. Many genomic variations are not on the genes and, are therefore not impactful. As there are several redundancies in the human genome, some of the variations on genes might not have an effect on the expressed phenotype either. However, when we look around, we see phenotypically diverse human characteristics, such as differences in appearance, personality, and physiology. Our genome gives rise to our traits. Some genomic variations cause diseases. Therefore, genomic data are tremendously important for many practical applications, ranging from screening for genetic diseases to forensic DNA fingerprinting. All of the applications of genomic data require the analysis of a genome for the presence of variation. This analysis can only be done with knowledge of other genomes, such as the average variation in a population and how common such variation is. Thus, the sharing of genomic data is the foundation of many studies related to genomic variation, especially those related to human health.

With the increase in promising solutions that personalized medicine offered to the individuals, and the decreasing cost of DNA sequencing technologies [4], clinical diagnostics are increasingly based on a patient's genetic make-up. The advances in next-generation sequencing technologies enabled an exponential increase in the number and size of available genomic datasets, and helped these data become

Responsible Genomic Data Sharing. https://doi.org/10.1016/B978-0-12-816197-5.00001-2

available to a wider group of audiences such as hospitals, research institutions, and individuals.

The sharing of genetic information gives rise to privacy concerns. Laws such as the Genetic Information Nondiscrimination Act have arien to protect individuals against biases based on genetics by insurance companies or employers in the case that their genetic data is obtained without their consent. Regardless of the potential monetary harm caused by unlimited sharing of genomic data, dignitary harm is also a factor when data is shared without consent [5]. In turn, individual privacy has become an important aspect of biomedical data-sharing regulations and policies, as well as research studies [6,7]. Establishing infrastructures for genomic data sharing with the consent of individuals is critical in advancing biomedical data science. Easy access to the data allows researchers to rapidly investigate health-related problems. It also permits speedy assessment of publishing tools and methods, and, in turn, increases reproducibility.

In this chapter, I will aim to convince you of the importance of data sharing for health and biomedical studies by laying out the facts from a researcher's perspective. I will discuss in detail the importance of data sharing in curing diseases and give examples from cases of rare diseases and cancer. I will also discuss the importance of statistical power in genome-wide association studies and how data sharing can help overcome problems related to sample size. I will emphasize the importance of data sharing from the perspectives of advancing basic research and discovery, the financial burden of generating a large amount of data by a single lab, and reproducibility in research. I will underscore the importance of developing general standards for data sharing to advancing science and the contribution of the well-curated databases to science.

2. Advancing research and scientific knowledge

Understanding human health and well-being starts, by definition, with examining the human body in health and sickness. Thus, the foundation of biomedical and healthcare studies relies on the data collected. The use of genetic information both in the clinical setting and in biomedical research is now playing a major role in unraveling the mysteries of many undiagnosed diseases. Genome sequencing technologies can help identify genetic variations in humans that cause or influence diseases ranging from Huntington disease to cancer. Functional genomics assays, such as RNA-Seq, can help us understand genetic activity that is different in disease and health. The integration of multiple assays has shown to be more useful in understanding human diseases [8]. However, it is extremely difficult and expensive to collect all types of genomic data at one site or institution. Therefore, data sharing across institutions and labs is essential for data integration. It is of key importance to share genomic data for the advancement of biomedical and healthcare studies. These studies are largely affected by the availability of datasets that face significant regulatory and institutional barriers due to rightful privacy concerns. In some cases, such barriers are not there but researchers prefer not to share data due to conflict of interest or not making data sharing a priority among other duties. The key is to come up with sensible and sensitive policies that enable extensive sharing of data and promoting the synthesis of data from different sources.

3. Importance of genomic data sharing in curing diseases

A disease is a medical condition associated with an abnormality that affects the function of an organism. A disease can be caused by external factors such as pathogens or stress or internal factors such as genetic make-up. The process of diagnosis and treatment of diseases start with physicians relating a patient's symptoms and test results to a known disease with characteristics that have been observed and learned from many patients in history. They then recommend treatments that were evaluated in the past and approved by clinical trials. However, these trials are often based on a small number of patients,whose experiences/make-ups may not be generalizable. The doom of the small sample size is even exacerbated when patients have rare diseases. Genomic data sharing can be crucial in identifying previously unobserved connections that could improve treatments for diseases with low-frequency mutations or with substantial genetic diversity.

3.1 Rare disease perspective

In the United States, a disease is characterized as rare if it affects fewer than 200,000 Americans at any given time. There are 30 million people in the United States and 350 million people in the world currently suffering from a rare disease. There are currently more than 6000 rare diseases identified and cataloged in the world. Only 5% of these diseases currently have FDA-approved treatments [9]. Moreover, 80% of rare diseases have been determined to have a genetic origin and 50% of these diseases affect children, pediatric cancer being one of them [10]. Rare genetic diseases are diseases that are caused by a mutation in the genome that is not observed frequently in the population. Next-generation genome sequencing provides a powerful tool to help patients with rare genetic diseases. Aggregation of such genomics data has the power to provide statistically sound diagnosis and targeted therapies. For example, the genetic causes of conditions such as Kabuki and Miller syndrome have recently been identified by comparative genomic studies using whole-exome and genome sequencing techniques [11,12].

One problem in understanding the biology behind these diseases is the diversity of symptoms even among patients who suffer from the same subtype of the disease. Misdiagnosis followed by mistreatment is common due to the lack of understanding and knowledge, hence the quality of patient life is extremely affected. There is a serious lack of scientific knowledge and quality information about rare diseases due to the lack of data. Diagnosis of such diseases is extremely difficult as it requires genetic data not only from the patient but also from many other patients with the same disease and data from a healthy population. As such, the development of treatments is extremely difficult due to a lack of statistical power to annotate the genomic mutations as disease-causing. It is a tremendous challenge for rare disease patients to receive accurate and timely diagnosis, and to get access to the correct treatments [13]. There are many different disease subtypes originating from different regions of the genome, which I will go over next. Hence, it is quite difficult for a single institution or even a country to come across a large number of patients with similar mutations. Therefore, not only curing rare diseases but even just understanding the biological mechanisms underlying these diseases is extremely dependent on accessing data from other patients in other institutions and countries. It is increasingly evident that there is great value in collecting and sharing genetic data on a worldwide scale in the context of rare diseases; the addition of one data point makes a substantial difference in sample size. Advancing next-generation genomics, along with better means for worldwide data sharing are the key components to achieving a better understanding of rare diseases. Due to these challenges, the infrastructure needs to be international and include a rigorous system for data sharing, aggregation, and fusion. There should be a feedback mechanism where researchers and clinicians are connected and can reliably exchange data.

Data access is extremely important in the diagnosis of rare diseases, but it is also important for curating better treatments. For example, by means of pharmacogenomics, one can look at how genetic variation affects the response of a patient to a drug [14]. Pharmacogenomics is a new field that emerged from the personalized medicine movement. The idea behind pharmacogenomics is to combine pharmacology and genetics to develop medications tailored to individuals' genetic make-up. The pharmacology field currently targets a generic patient, but drugs do not work in the same way for all individuals. Some individuals will benefit from the average drug, some will not respond and some will experience negative side effects. With the sequencing of thousands of patients, researchers are gaining insight into how genomic variation affects an individual's response to medications. Although the field of pharmacogenomics is quite new, the aim in the future is to develop personalized drugs to treat important diseases such as asthma or cancer. Therefore, access to knowledge of other patients' drug responses for rare genetic variation is likely valuable for curating effective therapeutics.

3.2 Cancer perspective

Cancer is one of the deadliest of all diseases. In short, cancer is a general name for a group of diseases that are driven by the uncontrolled growth of abnormal cells. These cells are often spread across different tissues and organs and cause death. The formation of malignant cells is known to be driven by several mutations in the genome that are accumulated throughout an individual's life. There are also mutations in the genome that are inherited and are known to determine the risk of acquiring these malignancies. According to the Cancer Facts and Figures 2019 report of the American Cancer Society [15], more than 1.7 million new cancer cases are expected to be diagnosed in 2019. About 606,880 Americans are expected to die of cancer in 2019, which translates to about 1660 deaths per day. Cancer is the second most common cause of death in the US.

The inheritance and accumulation of mutations in the genome is the main cause of many cancer types [16]. Some of these mutations hit important genes in the genome, which result in abnormal expression of those genes and ultimately result in uncontrolled growth of the cells with abnormal gene expression. The identification of important mutated genes that drive the formation of tumors is a major cancer research focus as these drivers are important for designing targeted therapies and for better understanding of the mutational processes that lead to oncogenesis. Although many of the cancer types are unfortunately commonly seen in the population, reaching statistical power for the count of the recurring mutations for designing targeted therapies is a problem. With the increase in technological developments and a decrease in the cost of DNA sequences, we now have genomic data from thousands of cancer patients [17]. However, one problem with the cancer studies is that we now know that the distribution and landscape of the mutations that directly or indirectly play a role in driving the formation of malignancies largely vary from patient to patient. The somatic mutation frequency distribution in different cancers also varies from one cancer to another. Moreover, within a single cancer type, the frequency of mutations differ from patient to patient. This creates difficulties identifying genes that are driving the cancer event with high statistical confidence. Despite the fact that we now have genetic data from thousands of cancer patients, the sample size of observing a recurrent mutation on a gene is still quite low. Moreover, the wide distribution of mutation frequency also means that there are unidentified genes that could potentially be targeted in therapies. All these problems can only be solved if data from many sources can be put together.

Today, not only researchers in academic institutions but also healthcare providers are shifting toward data-sharing practices with the understanding that data sharing will transform and help discover new ways to treat cancer. There is also another shift in the cancer healthcare community, which is the use of precision medicine with treatments tailored to patients and their tumors. This precision medicine approach is based on the patient's clinical, molecular, and genetics data, as well as data from the outcome of treatments and response to drugs. There is a massive effort to gather such data from patients in many healthcare systems. However, the current situation is that such data are confined within the particular hospital database and not shared widely with the research community. There is currently not a unified way of storing and sharing such data or any standards to guide the format of the collected data [18].

Increasing amounts of accessible data lead to improvement in patient care, especially if a precision medicine approach is taken. For a doctor, knowledge of how certain targeted therapeutics affected by other patients with a similar genetic make-up is extremely important for making treatment decisions as well as discovering new ways of treating diseases. This is, of course, only possible if such data are shared across institutions. The best therapies for a cancer patient with a rare mutation can be decided at the point of care by using data and experiences from thousands of other patients across institutions and hospitals.

The National Cancer Institute Genomic Data Commons (GDC) is a tremendous step toward democratizing and sharing high-throughput sequencing of cancer genomes. It is a system that many cancer biologists and oncologists can easily access and permits searches of thousands of cancer mutations. Many healthcare organizations are now contributing data to GDC to aid cancer research. Among those organizations are the Multiple Myeloma Research Foundation, the Knight Cancer Institute at Oregon Health and Science University, and the Dana-Farber Cancer Institute. GDC is an open-resource software platform that provides retrieval of the data, using cloud-based computing infrastructure with commonly used software and tools. CDC was launched in 2016 and today, it has over 2.7 PB of genomic and associated clinical data and is being used by more than 100,000 researchers every year. Among the tools that GDC provides are Jupyter notebooks, RStudio notebooks, and genome analysis software such as GATK [19,20], BWA [21], and many more. Three cloud-based computing platforms are associated with CDC: Broads FireCloud, the Seven Bridges Genomics Cancer Genomics Cloud, and ISBs Cancer Genomics Cloud. In these platforms, users can easily access the genomic data and perform common computations such as sequence alignment or variant calling with a few clicks by using publicly available tools [22,23]. The data collected from these multiple institutions will help researchers to investigate causal mutations in cancer, find out therapeutic targets or even design software and tools that work better with cancer karyotypes.

The patients at the intersection of both rare diseases and cancer will also greatly benefit from data sharing. The US Department of Health and Human Services and National Institutes of Health (NIH) have recently announced a special focus on pediatric cancer [10]. Every year, approximately 15,300 children are diagnosed with pediatric cancer. Childhood cancer is the number one cause of death by disease in children regardless of age, ethnicity, socioeconomic class. The survival rate is much lower compared to other cancer types. On top of that, the number of childhood cancer diagnoses has not decreased in almost 20 years. Every day, around 43 children are diagnosed with childhood cancer and 12% of these diagnosed children do not survive. There is no difference in statistics between ethnic groups, genders or economic groups. The average age of diagnosis is 6 and more than 40,000 children are undergoing treatment each year. Moreover, 60% of these children who survive cancer suffer late-effects such as infertility, secondary cancers, and heart failure. Although the United States has only 50% of the world's pediatric cancer cases, 1 in 530 adults ages 20–39 in the United States are adult survivors of childhood cancer [24].

Due to the rarity of childhood cancer and the broad spectrum of rare mutations, it is extremely difficult to obtain statistical power to find genetic causes of pediatric cancer. Therefore, to further the understanding of these cancers and their particular subtypes, researchers and oncologists must be able to have access data from similar cases, which are likely available in a different institute or even in a different country. Assembling large cohorts to achieve statistical power is not only a problem from a diagnostic perspective but also a barrier to clinical trials. Cancer immunotherapy is an effective way of battling cancer, and is based on artificial stimulation of the immune system to attack cancer cells and kill them. It takes advantage of the fact that tumor cells have antigens on their cell surface that can be detected by antibodies. There are various genetic indicators that predict the probability of effectiveness of immunotherapy treatment. Microsatellite Instability and Tumor Mutational Burden were approved by the FDA in 2008 to be used to predict whether immunotherapy will be effective. Tumor Mutational Burden is the number of mutations within a targeted genetic region in the DNA of the tumor cell. A big hurdle in applying immunotherapy to pediatric cancer is the observation of the low number of tumor mutational burden as only a few subtypes of pediatric cancer have recurrent mutations that can be targeted by immunotherapy. On the other hand, there is a large differential gene expression indicating a massive genetic misregulation and can only be understood if the information is shared. A combination of transcriptomics and genetic data will likely give the full picture of misregulation in pediatric cancer. However, a single institution or hospital or a country may not have the financial abilities to perform all of these tests at once. This is when data sharing becomes extremely critical. Sharing pediatric cancer data is extremely challenging due to additional privacy issues that come with underage patients [10]. Another challenge in deriving pediatric cancer data is that underage individuals cannot be included in studies that require informed consent [10].

The Pediatric Cancer Data Commons (PCDC) is another initiative to curate high quality, unified childhood cancer data for researchers. PCDC works with researchers around the globe to help them easily share and integrate data, query the database and request and analyze data on demand. PCDC distinguishes itself from other data commons with an expertise in international privacy laws, which means that they are able to understand the international data-sharing agreements, which, in turn, allows them to collect data from more patients and more different backgrounds. Another focus of PCDC is to establish common data standards to achieve a unified resource. In addition to PCDC, there are other efforts to collect and harmonize childhood cancer data from St. Jude's Children Hospital, Children's Tumor Foundation efforts at Sage and Kids First initiative.

3.3 Genome-wide association studies perspective

A genome-wide association study (GWAS) is based on observations of a genome-wide set of genetic variants in different individuals to examine if any of these genetic variants are associated with a trait [25]. These studies typically study the effect of single-nucleotide variants on traits and human diseases but recently, they have been extended to other genetic variants such as structural variants. They have also been extended to other organisms for ecology and evolutionary biology studies.

The core of a GWAS study involves scanning markers across many individuals and aggregating the data over large cohorts. Therefore, such studies are particularly useful in finding genetic variants that contribute to common diseases such as diabetes, asthma or heart diseases, and common human traits.

One can envision that the impact of such studies on human health will be accelerated with the initiation of the personalized medicine approach. Today, even with direct-to-consumer genetic testing, it is possible to tailor diagnosis to individuals based on their genetic variants. Risk assessment with certain diseases is possible by scanning through hundreds of GWA studies that were done on thousands of individuals.

GWA studies are conducted using two groups of participants. Group 1 includes individuals with the disease or trait being studied and group 2 includes individuals without the disease or trait studied. DNA from each participant is obtained and scanned for known genetic variants usually using a gene-chip approach. These chips target strategically selected markers of genetic variants. If a variant is significantly present in the cohort with disease or trait, then that genetic variant can be linked to the disease or the cohort. Similarly, if a variant is significantly present in the normal cohort, then the absence of such variants can be linked to the disease or trait.

One can realize quickly that statistical power is an important property of GWAS. For example, as the number of individuals in the study increases, the number of a true marker of genetic variants might increase and false positives might disappear. There might be GWA studies conducted across different institutions for the same genetic traits of diseases. It is extremely useful to aggregate these studies to increase the statistical significance of the findings and enable the discovery of new genetic variants associated with diseases. Therefore, it is extremely important to share GWAS data across research institutions. It is also important to share such data across hospitals and healthcare providers for better diagnostic and risk assessment of patients through personalized medicine.

GWA studies are available through the National Center for Biotechnology Information (NCBI) as a part of NIH's National Library of Medicine for the research community. NCBI's website Database of Genotype and Phenotype [26] (dbGap) hosts a variety of GWASs on different diseases and conditions. Following the seminal work by Homer et al. [27], summary results of these datasets were put behind a special access firewall; however, as of December 2018, a new data access policy by NIH gave the public access to the summary statistics while the details of the full study are accessible through controlled dbGaP access.

Similar to many other studies that are based on a cohort of individuals, GWA studies also suffer from several limitations that require a well-thought out study setup. One of the limitations that these studies face is insufficient sample size, which in turn creates problems related to multiple testing and control for population stratification. One way to overcome this limitation is to collect and share data for better statistical power.

Many GWAS have been performed for psychiatric disorders. Although these studies provide a wealth of information about the genetics of these diseases, the number of genetic variants obtained from these studies explain the variation in the disease and the heritability poorly [28]. One of the reasons for such poor performance is the lack of statistical power as the number of study individuals is limited and hence the resulting number of genetic variants that are captured is low. As psychiatric disorders are often seen as stigmatizing phenotypes, it is quite possible that many institutions keep the collected data to themselves. However, statistical power can be obtained by combining studies from multiple sites. This will only be possible through standardized and responsible sharing of data between institutions.

4. Impact of large-scale data sharing on basic research and discovery

The NIH and the Department of Energy along with partners from all over the world aimed to sequence the complete genome of humans in 1990 in an effort called the Human Genome Project [29]. This public effort aimed at enabling researchers to investigate the genetic factors in human disease by providing a powerful resource, which eventually paved the way for new diagnosis, treatment, and prevention strategies. With the completion of the Human Genome Project, many revolutionary biotechnological innovations and large-scale projects were launched. Because of this concerted and freely and publicly available resource, researchers discovered more than 1800 disease genes. While in the past it could take years to find the heritability of a variation in a gene causing a disease, today we can investigate it in a matter of hours or days because of the availability of the human genome. Researchers developed more than 2000 genetic tests for different human conditions and these tests allowed us to investigate the genetic risks for different diseases and shaped today's healthcare system with better diagnostic tools [29].

After the completion of the Human Genome Project, a variety of large international consortium projects have launched with an open data-sharing policy. The data obtained from these studies are still largely being used in basic research and have advanced our understanding of human health and disease. In the following sections, I will describe these projects with an emphasis on their impact on basic research because of publicly shared datasets provided by these initiatives.

4.1 The International HapMap Project

One of the major biotechnological projects that has launched with the completion of the Human Genome Project is the International HapMap project that started in 2005 [2]. This project aimed at cataloging the common genetic variation in the human genome. The goal was to develop a haplotype map of the human genome to provide a resource for researchers to find disease associating genes and their response to drugs. Importantly, the data produced are now freely available in public databases. The third phase of the HapMap project was published in 2010, spanning data from 11 global populations. HapMap data have been used for studies in finding genetic factorsfor common diseases, have generated multiple impactful research results, and have changed the way we think about genetic risks and diseases. This project was a collaboration among researchers at academic centers, nonprofit biomedical research groups, and private companies in Japan, the United Kingdom, Canada, China, Nigeria, and the United States. The impact of the HapMap project through publicly available datasets is massive. International HapMap samples are composed of lymphoblastoid cell lines (LCLs) derived from individuals from different world populations. These samples were shown to be extremely helpful for developing targeted therapeutics. This project was particularly useful because samples were derived from different major geographical populations, allowing for a detailed study of population stratification.

4.2 1000 Genomes Project

1000 Genomes Project has launched with the aim of finding most genetic variants with frequencies of at least 1% in the populations studied using some of the samples from the HapMap project [3]. With the advancement of next-generation sequencing technologies, the 1000 Genomes Project was the first to use sequencing technology to provide a comprehensive resource on human genetic variation. Data was made publicly available to the worldwide scientific community as the project continues. At the time of the launch of the study, next-generation sequencing technologies were still such that it was quite expensive to sequence the samples deeply. However, after sequencing many samples and aggregating them, the project developed ways to discover a comprehensive map of the genetic variation across human populations. Moreover, 2504 samples were combined to allow accurate genotyping of the samples combined with the genotype imputation approach at the last phase of the project.

Furthermore, the 1000 Genomes Project focused on samples with broad consent, which are open-access worldwide. Not only the derived genotypes and sequencing files but also LCLs and DNA samples derived from 1000 Genomes individuals are available through NIH repositories. In 2008, the project announced the release of the first four samples. The data from these samples were used in 100s of biomedical studies.

The release of 1000 Genomes data did not only lead to biological discoveries but also improvements in information technology and bioinformatics. Owing to the huge amount of data and type of data, many bioinformatics algorithms from genotyping to sequence alignment tools were developed to better understand the collected data. 1000 Genomes project also developed a web-based browser, showing an example of infrastructure for the dissemination and query of such large datasets for the scientific community and other large consortia.

The open and public release of 1000 Genomes data enabled the shaping of other useful resources. For example, a pharmacogene database was released to utilize the 1000 Genomes data [30,31]. This database is composed of genotypes of pharmacogenetic candidate genes by the Very Important Pharmacogenes project of the Pharmacogenetics Knowledge Base (PharmGKB) on 35 HapMap CEU and 26 HapMap YRI samples (April 2009) [30,32].

4.3 The Cancer Genome Atlas project

Much work is still needed before we can unravel the mysteries of human genome variation and how it affects our health. As a result, another large initiative has recently started aiming to identify the genetic causes of major cancer types, called The Cancer Genome Atlas (TCGA) [17]. With over $300 million in total funding, TCGA's mission was to catalog the genomic changes underlying multiple cancer types. TCGA was initiated in December 2005 and initially focused on three different cancer types: brain, lung, and ovarian. Today, it contains around 33 different tumor types including 10 rare cancers with over 2.5 petabytes of data shared publicly.

With the reduction in the cost of DNA sequencing and collaboration between around 20 different institutions, researchers provided and shared a vast amount of data including exome and whole-genome sequencing, gene-expression profiling, copy-number variation profiling, and single-nucleotide polymorphism genotyping, gene expression, and more. Not only the data but the physical samples from cancer patients were shared by the source institutions across the world.

TCGA is an immeasurable resource for all the scientists around the globe, with data from over 11,000 patients and contributions from thousands of researchers. It changed our understanding of cancer and the diagnosis and clinical treatments. The impact of this resource can be seen in many fields, such as the development of new biotechnology tools and new software in computational biology.

Because of TCGA, we now know that there are other types of genome alterations such as copy number changes and structural variants beyond single nucleotide changes that affect the disease status of a genome [17]. TCGA resource goes beyond genetics and with the immense amount of data it contains, researchers are now able to understand the aberrations in gene expression, epigenetic and protein expression, and structure in cancer cells [17]. The data helped researchers from diverse fields such as metagenomics and immunology and researchers studying other diseases.

One of the main contributions of TCGA was in the field of computational biology. The projects involving TCGA data range from somatic and germline SNP calling tools to regulatory network construction and image processing. TCGA data are hosted by the genomic data commons, which allows not only the query and access of data but also now hosts a cloud-based genomics application allowing researchers to conduct analysis on demand.

The TCGA consortium is a member of the International Cancer Genomics Consortium (ICGC). ICGC's mission is the global analysis of all possible tumor types seen in the world. The ICGC PanCancer Analysis of Whole Genomes (PCAWG) combined a vast amount of whole-genome sequencing data from around the globe and TCGA provided half of the PCAWG data.

4.4 Encyclopedia of DNA Elements Project

ENCODE, the encyclopedia of DNA elements, is a public research consortium supported by The National Human Genome Research Institute (NHGRI). The mission of the consortium is to identify the functional elements in the human and mouse genome beyond coding sequences. The tremendous amount of epigenomics data ENCODE produced are immediately released at the ENCODE data portal and freely accessible to the community. The ENCODE data are organized into three levels with the first level being the raw data and the second and third levels being the two levels of annotations: integrative-level annotations, with the registry of candidate *cis*-regulatory elements and the ground-level annotations that are directly derived from the experimental raw data. ENCODE portal hosts data from other consortium and projects such as modENCODE [33], RoadMap Epigenomics and Genomics of Gene Regulation [34], and EN-TEx, which includes functional genomics assays of four different individuals selected from Genotype-Tissue Expression (GTEx) samples.

The first phase of the ENCODE project started in 2003 [36]. It is currently in its fourth phase. Along with the ENCODE effort, the data production and technology development initiatives have been funded by NHGRI since 2003. In addition to the well-studied cell lines, the ENCODE project also uses primary cells and tissues that have been consented for the unrestricted sharing of genomic data. This allows the community to access all of the data generated by the Consortium freely and maximizes the utility.

ENCODE is a comprehensive resource that allows the scientific community to understand the function of the human genome, and will lead to a better understanding of human health. According to the ENCODE data portal, as of May 2019, the consortium published around 972 publications on the human genome alone. A total of 1949 publications involved ENCODE data from labs with no ENCODE funding; 721 of them relate to human diseases, 777 of them are basic biology publications, and 279 of them are software tools. These numbers show the importance of this consortium in the community and also highlight the importance of unrestricted access to data in accelerating biomedical research.

ENCODE project is currently in its fourth phase, which focuses on studying a broader diversity of biological samples, especially the ones with disease association. The fourth phase is also exploring new omics assays that were not previously used in ENCODE. As with the previous phases, the data from the fourth phase will be available through GEO and ENCODE data portal with no restrictions.

4.5 Genotype-Tissue Expression Project

The GTEx project was launched in 2010 by the NIH to investigate how variation in the human genome affects tissue expression. GTEx created a massive resource that includes whole-genome, whole-exome, and RNA sequencing of 53 healthy tissues from nearly a thousand consenting individuals. This resource does not only contain an open-access databank but also provides a biobank that researchers can access and use for future studies.

It is important to point out that relatives of nearly 1000 deceased individuals generously donated these tissues for research scientists to understand how human genes work so that better ways to prevent, diagnose, and treatment for diseases can be found in the future.

The GTEx Portal provides open access to data including gene expression, quantitative trait locus, and histology images. The main idea behind the GTEx project is to find correlations between tissue-specific gene expression and genotype. This allows researchers to find target regions in the genome that affect the gene expression the most. As a result, these regions will be the candidate regions for disease susceptibility.

GTEx project has already provided an immense amount of data to the community to study human gene expression and its relationship to genetic variation. As we mentioned earlier, GWAS is a powerful tool to identify the genetic variation that is associated with common human diseases. However, most human genetic variation is outside of the protein-coding regions of the genome. This makes it difficult to interpret the GWAS results and understand the mechanism of regulation. GTEx aimed at correlating the genetic variation directly with gene expression, rather than a high-level phenotype such as asthma or heart disease like in the case of GWAS. This creates a valuable resource for researchers to understand the mechanism of gene regulation through genome variation. GTEx consortium published over 20 publications. The flagship publication in 2013 that was published in *Nature Genetics* [37] has been cited more than 1500 times so far.

5. Reproducibility

In a poll launched by Nature in 2016, 70% of 1500 scientists claimed that they failed to reproduce at least one other publication's experiment. This reproducibility crisis is known to affect life sciences more than other fields. This is unfortunate considering life sciences deal with human health and well-being. Replication in sciences that relate to human health is important due to a direct relationship between science and human health, disease discoverie, and development of better diagnosis and treatment tools. Although access to data is not the only reason causing replication problems, in computational sciences, encouraging access to data will increase reproducibility because scientists will be able to test methods and software with the exact same data reported. To this end, sharing the genomic data that many tools are developed to analyze is extremely important for the rapid assessment of these tools.

6. Patient/public perspective

As scientists, we believe that we are using data for the greater good and therefore data sharing is important. However, this scientific perspective often overlooks the perspectives of patients and the public. Medical doctors believe that many patients do not think about data sharing as much as we think they do. When genomic data sharing is mentioned, patients are typically most concerned about data being shared with entities that profit from them. They often assume that their data will be used in the best possible way to help other patients [35].

When it comes to genomic data sharing, privacy must have the highest priority as DNA is the blueprint of one's being and contains extensive identifying information not only about the patient but also about the relatives of the patient. This is where informed consent comes into play, in which a patient is informed about the risks of sharing their data. Whether or not the patient clearly understands what the informed means, patients are increasingly changing their ways of thinking about data sharing. According to a poll by Global Alliance for Genomics & Health presented at their annual meeting in 2018, patients are more inclined to share their genomics data if the data are being used by nonprofit institutions for research purposes. It appears that patients are more hesitant about data sharing and participating in studies if data are shared by for-profit commercial entities. This begs the question of why the direct to consumer (DTC) companies such as 23andme.com and ancestry.com are gaining more and more customers. It could be that the public views these DTC companies as service providers rather than commercial entities sharing their data.

The public's views on genomic data sharing also show variation. Patients or individuals who are extremely sick do not seem to think about who owns and controls their genomic data. They are interested instead on sharing their data quickly and with multiple institutions in case one can find novel ways of treating their medical condition. On the other hand, healthy individuals tend to think more about their data ownership, and the potential for leaking their identifying information. This difference in values between healthy and sick patient groups presents a tricky situation as far as data sharing policies go; it becomes difficult to establish a unified framework that will satisfy that the concerns of both groups.

There are not enough social studies investigating the patient's view of genomic data sharing and informed consent. Among all the "big data" genomics consortia, GTEx was the only project that performed social studies on consultation and research into the ethical, legal, and social issues raised by the biomedical research and data storage and sharing.

Acknowledgments

I would like to thank my postdoctoral advisor Prof. Mark Gerstein for valuable inputs and opinions on privacy issues related to genomics data sharing. I also would like to thank Prof. Casey Greene for sharing his valuable insights on genomic data commons, rare disease data as well as data-sharing standards. Many thanks to Charlotte Brannon for helpful edits and fruitful discussions.

References

[1] International Human Genome Sequencing Consortium. Initial sequencing and analysis of the human genome. Nature 2001;409(6822):860–921.

[2] International HapMap Consortium. The international HapMap project. Nature 2003; 426(6968):789–96.
[3] The 1000 Genomes Project Consortium. A map of human genome variation from population scale sequencing. Nature 2010;467(7319):1061–73.
[4] Sboner A, Mu X, Greenbaum D, Auerbach RK, Gerstein MB. The real cost of sequencing: higher than you think! Genome Biology 2011;12(8):125.
[5] Joly Y, Feze IN, Song L, Knoppers BM. Comparative approaches to genetic discrimination: chasing shadows? Trends in Genetics 2017;33(5):299–302.
[6] Joly Y, Dyke SOM, Knoppers BM, Pastinen T. Are data sharing and privacy protection mutually exclusive? Cell 2016;167(5):1150–4.
[7] Erlich Y, Narayanan A. Routes for breaching and protecting genetic privacy. Nature Reviews Genetics 2014;15(6):409–21.
[8] PsychENCODE Consortium. Revealing the brain's molecular architecture. Science 2018;362(6420):1262–3.
[9] National Institute of Health FAQs about rare diseases https://rarediseases.info.nih.gov/diseases/pages/31/faqs-about-rare-diseases.
[10] Vaske OM, Haussler D. Data sharing for pediatric cancers. Science 2019;363(6432):1125.
[11] Ionita-Laza I, et al. Finding disease variants in Mendelian disorders by using sequence data: methods and applications. The American Journal of Human Genetics 2011;89(6):701–12.
[12] Ku CS, Naidoo N, Pawitan Y. Revisiting Mendelian disorders through exome sequencing. Human Genetics 2011;129(4):351–70.
[13] Austin CP, et al. Future of rare diseases research 2017–2027: an IRDiRC perspective. Clinical and Translational Science 2018;11(1):21–7.
[14] Weng L, Zhang L, Peng Y, Huang RS. Pharmacogenetics and pharmacogenomics: a bridge to individualized cancer therapy. Pharmacogenomics 2013;14(3):315–24.
[15] American Cancer Society Cancer Facts and Figures 2019. https://www.cancer.org/research/cancer-facts-statistics/all-cancer-facts-figures/cancer- facts-figures-2019.html.
[16] Stratton MR, Campbell PJ, Futreal PA. The cancer genome. Nature 2009;458(7239):719–24.
[17] Cancer Genome Atlas Research Network, Weinstein JN, Collisson EA, Mills GB, Shaw KR, Ozenberger BA, Ellrott K, Shmulevich I, Sander C, Stuart JM. The cancer genome Atlas pan-cancer analysis project. Nature Genetics 2013;45(10):1113–20.
[18] Clinical Cancer Genome Task Team of the Global Alliance for Genomics and Health, et al. Sharing clinical and genomic data on cancer - the need for global solutions. New England Journal of Medicine 2017;376(21):2006–9.
[19] DePristo M, Banks E, Poplin R, Garimella K, Maguire J, Hartl C, Philippakis A, del Angel G, Rivas MA, Hanna M, McKenna A, Fennell T, Kernytsky A, Sivachenko A, Cibulskis K, Gabriel S, Altshuler D, Daly M. A framework for variation discovery and genotyping using next-generation DNA sequencing data. Nature Genetics 2011; 43(5):491–8.
[20] Van der Auwera GA, Carneiro M, Hartl C, Poplin R, del Angel G, Levy-Moonshine A, Jordan T, Shakir K, Roazen D, Thibault J, Banks E, Garimella K, Altshuler D, Gabriel S, DePristo M. From FastQ data to high-confidence variant calls: the genome analysis toolkit best practices pipeline. Current Protocols in Bioinformatics 2013;43. 11.10.1–11.10.33.

[21] Li H, Durbin R. Fast and accurate short read alignment with Burrows-Wheeler transform. Bioinformatics 2009;25:1754–60.

[22] Jensen MA, et al. The NCI Genomic Data Commons as an engine for precision medicine. Blood 2017;130(4):453–9.

[23] Grossman RL. Data lakes, clouds, and commons: a review of platforms for analyzing and sharing genomic data. Trends in Genetics 2019;35(3):223–34.

[24] Curesearch For Children's Cancer Childhood Cancer Statistics. https://curesearch.org/Childhood-Cancer-Statistics.

[25] Manolio TA. Genome-wide association studies and assessment of the risk of disease. New England Journal of Medicine 2010;363(2):166176.

[26] Mailman MD, Feolo M, Jin Y, Kimura M, Tryka K, Bagoutdinov R, Hao L, Kiang A, Paschall J, Phan L, Popova N, Pretel S, Ziyabari L, Lee M, Shao Y, Wang ZY, Sirotkin K, Ward M, Kholodov M, Zbicz K, Beck J, Kimelman M, Shevelev S, Preuss D, Yaschenko E, Graeff A, Ostell J, Sherry ST. The NCBI dbGaP database of genotypes and phenotypes. Nature Genetics 2007;39(10):1181–6.

[27] Homer N, Szelinger S, Redman M, Duggan D, Tembe W, Muehling J, Pearson JV, Stephan DA, Nelson SF, Craig DW. Resolving individuals contributing trace amounts of DNA to highly complex mixtures using high-density SNP genotyping microarrays. PLoS Genetics 2008;4(8):e1000167.

[28] Collins AL, Sullivan PF. Genome-wide association studies in psychiatry: what have we learned? British Journal of Psychiatry 2013;202(1):1–4.

[29] NIH Human Genome Project https://report.nih.gov/NIHfactsheets/ViewFactSheet.aspx?csid=45.

[30] zheng W, Dolen EM. Impact of the 1000 Genomes Project on the next wave of pharmacogenomic discovery. Pharmacogenomics 2010;11(2):249–56.

[31] Gamazon ER, Zhang W, Huang RS, Dolan ME, Cox NJ. A pharmacogene database enhanced by the 1000 Genomes Project. Pharmacogenetics and Genomics 2009; 19(10):829832.

[32] Klein TE, Chang JT, Cho MK, et al. Integrating genotype and phenotype information: an overview of the PharmGKB project. The Pharmacogenomics Journal 2001;1(3): 167170.

[33] Brown JB, Celniker SE. Lessons from modENCODE. Annual Review of Genomics and Human Genetics 2015;16:31–53.

[34] Roadmap Epigenomics Consortium, et al. Integrative analysis of 111 reference human epigenomes. Nature 2015;518(7539):317–30.

[35] Haug CJ. Whose data are they anyway? Can a patient perspective advance the data-sharing debate? New England Journal of Medicine 2017;376:2203–5.

[36] ENCODE Project Consortium. An integrated encyclopedia of DNA elements in the human genome. Nature 2012;489(7414):57–74.

[37] The GTEx Consortium. The genotype-tissue expression (GTEx) project. Nature Genetics 2013;45(6):580–5.

CHAPTER

Genomic data access policy models

2

Stephanie O.M. Dyke, PhD

Academic Associate, McGill Centre for Integrative Neuroscience, Montreal Neurological Institute, Department of Neurology & Neurosurgery, Faculty of Medicine, McGill University, Montreal, QC, Canada

This chapter provides an overview of the current state of data access policy models for sharing genomic research data and discusses the evolution of data-sharing policy in science as well as its ethical—legal underpinnings.

1. Data-sharing policy developments

Data-sharing policy has today become widespread across academic research as a result of policy developments over the past 25 years. The genomics community has played a leading role in driving data sharing across the life sciences.

In the 1990s, funders and other leaders in the genetics community realized that the Human Genome Project (HGP), an international collaboration to produce the first human genome map [1], would only be successful if the genetic sequence data produced by all participating centers were shared both very quickly, to enable immediate collaboration between teams around the globe, and publicly [2]. Indeed, a race to claim intellectual property rights on the human genome sequence had rapidly begun [3]. A formal open-access data-sharing policy was therefore widely adopted by the genomics research community: the so-called Bermuda Principles that guided the HGP [4]. In light of the success of the HGP, similar data-sharing principles were reaffirmed several years later in the Fort Lauderdale Report [5] (2003) to prepare for the next wave of large-scale community resource projects in genetics, such as the Mouse Genome Project [6] and HapMap Consortium [7].

In human genetics, the human genome map produced by the HGP led to the genotyping of large groups of individuals to study genetic variation. Genomic data-sharing policies, therefore, had to account for the ethical—legal risks inherent in sharing individual-level genetic data, as this was widely considered to raise serious privacy concerns [8]. This was further compounded with the advent of next-generation sequencing technologies in the mid-2000s, which enabled the sequencing of individual whole genomes. The Toronto Statement on prepublication data sharing, published in 2009, included the consensus statement that "for clinical and genomic data that are associated with a unique, but not directly identifiable

Responsible Genomic Data Sharing. https://doi.org/10.1016/B978-0-12-816197-5.00002-4

individual, access may be restricted" [9]. This marked the move from open access toward restricted data access policy models (controlled access and more recently registered access), whereby data were no longer published openly on the World Wide Web, but only made available under certain conditions.

At the same time, data-sharing policies have gradually expanded to cover not only large-scale community resource projects, but most data from genomics projects that could be used more broadly by the research community for further analysis, for integration into meta-analyses, or for use as control data and in methods development [10]. Along with its positive impact on scientific progress through greater efficiency and transparency, the benefits of data-sharing policy both for academic integrity and for research translation and uptake strategies rapidly gained recognition [11,12]. There are currently overarching data-sharing policies for most public and charitable genomic research funding in the United States, the United Kingdom, and Canada [13–15], with further policy plans extending to data sharing in other areas of the life sciences as well as most other fields of academic research [16,17]. Many scientific journals have also adopted open-access publishing models due to funders' growing commitments in this respect and have required data sharing as a condition of publication [18–21]. These new funding and editorial requirements have also served to advance data-sharing practices.

The timing of actual data release, at the time of publication or before publication (prepublication release), as well as the specific conditions of data access and reuse have been, and continue to be, the focus of much attention and debate. This chapter focuses on the conditions of data access and reuse, so-called data access policy models.

2. Open-access policy model

As evident from the name open access, there are no restrictions or conditions imposed on access to the data in this model. This model has evolved with close ties to the academic drive for open-access publishing. Indeed, science funders' publication policies often present access to the data underlying published reports of research (i.e., data sharing) as the next step in open-access publishing [22].

In current open-access publishing models, academic manuscripts are not held behind journal subscription paywalls. They are publicly available, either from the journal website or from one of the public manuscript archives, such as PubMed Central. However, the exact terms and conditions of access to open-access publications vary somewhat from journal to journal, with different licensing options guiding the terms of reuse of the contents of manuscripts [23]. This is particularly critical for efforts to text-mine the content of the scientific literature, for example, to search for genes of interest across published studies and identify related pathways or phenotypes. Furthermore, in addition to licensing options that can be more or less restrictive, many journals offering open-access options also use open-access embargo periods, whereby they retain subscription-only access for a period of time, thus delaying open access of scientific publications. These embargo periods usually range from 6 to 12 months from the time of publication, as permitted by funders' open-access policies. Research funders in Europe recently announced they will no longer support open-access embargo periods (Plan S [24]).

Ideally, once a scientific report is published, the data used for the research analysis described in that publication should also be available. Open-access data, therefore, typically refer to data that are published openly, without any restrictions on their use, and that is freely available from databases and websites through the internet. In this respect, major public repositories were established at the National Center for Biotechnology Information (NCBI) in the United States, the European Bioinformatics Institute (EMBL-EBI), and the DNA Data Bank of Japan (DDBJ), to preserve and make open-access genomic data, such as the HGP data, widely available to the community [25]. Nevertheless, in practice, data with some access conditions are also widely considered to be shared under the open-access model. For example, data with terms of use requiring the acknowledgment of those who produced and shared the data or data shared with a "publication moratorium." In contrast to the open-access embargoes used by journals in open-access publishing (described earlier), publication moratoria are restrictions on any other scientist publishing the results of data analyses using shared data until a specific time limit has passed, or until the team sharing the data has itself published its first analysis of the data.

There are major advantages to the open-access model of data sharing. The rapid and unfettered access it provides encourages large numbers of investigators, including those in nonacademic settings, to access and use data shared this way. No resources, be they time or other, need to be spent on application procedures. Furthermore, with the exception of database copyright in some jurisdictions, which could limit certain types of data reuse, open-access genomic data are usually considered to be in the public domain, thereby enabling widespread commercial applications and development.

There are however two major drawbacks to the model. First, there cannot be any control over who is using data or what it is used for. This is of particular concern for data from individuals whose consent to the research study would limit its use to a specific area of investigation, such as health research. Second, there can, in some cases, be a risk of loss of privacy. Loss of privacy could occur, for example, through reidentification of shared data, whereby a research participant's health or other information used in research would no longer be anonymous (see Chapters 3, 4). When surveyed about their perceptions of data-sharing risk, research participants indicate concern about potential discrimination and stigmatization resulting from the misuse or reidentification of shared research data, especially for employment and insurance opportunities [26,27]. These risks have led to privacy and data protection, as well as anti-discrimination, laws and regulations that govern the sharing and use of health-related data (either from healthcare or produced specifically for research) [28–32]. Together, these ethical considerations and research regulations call for additional protections on data access and use to limit the risk to participants when sharing extensive, individual genomic data from research.

However, if today much genomic and health-related data produced and used in health research is not suitable for open-access sharing, it remains the model of choice for non-human genomic data. Open access is also widely used to share aggregate or summary human data that are not associated with particularly sensitive phenotypes (e.g., gnomAD browser [33]), as well as for sharing "reference" human genomic data, such as disease variants and their clinical significance (BRCA Exchange [34]).

3. Controlled-access policy model

The open-access data-sharing model that had been the norm in the early stages of genomic research could not any longer be satisfactory in the face of the risks incurred when sharing detailed individual genomic data on a wide scale. A major departure from that model was therefore necessary. However, scientists, funders, and journals were still keen to see the broad sharing of data, but it was clear that such data could not be shared publicly [35]. New data-sharing policies, therefore, emerged to enable "controlled access" (sometimes called "managed access") sharing, that is, sharing data with approved researchers for approved research purposes [36,37]. Secure public archives were established to store and distribute data requiring such access policy at the major public genomic repositories in the United States and Europe: the database of Genotypes and Phenotypes dbGaP at NCBI [38] and the European Genome-phenome Archive EGA at EBI [39]. Importantly, these archives had to allow the removal of individual datasets if a participant were to withdraw their consent. Although controlled data access policies and procedures differed somewhat between projects in different places [40,41], in the United States they were standardized for all data shared *via* dbGaP and centralized through a network of data access committees (DACs) supported by the National Institutes of Health (NIH) [42].

An international review of controlled-access procedures showed nine criteria that researchers applying for access to controlled-access datasets must typically satisfy (see Fig. 2.1) [43]. Applicants must be affiliated with a recognized scientific institution, whether public or private, commercial or non-commercial. They must be qualified to undertake the proposed data analysis. The analysis itself must be compatible with the data provider's objectives. The study must be compatible with consent requirements and the data requested must be necessary for the purposes of the analysis. Furthermore, all applicable ethical obligations, such as gaining local ethics approval for the study, must be met. Additionally, the scientific merit of the proposed analysis may be evaluated, as well as the adequacy of the applicant's data security policies. Finally, the applicant's institution is required to sign an access agreement to govern the research use of controlled-access data. These conditions of access are usually assessed by a DAC or data access office (DAO).

Within the bounds of their restrictions on who may access data and for what purposes, controlled-access policies generally attempt to maintain the "public domain"

FIGURE 2.1

The main steps in the review of applications under controlled-access data access policy. To access data, researchers must satisfy these conditions of access.

benefits of open-access sharing through the intellectual property terms of data access agreements. Standard clauses typically prohibit using intellectual property protection in ways that would prevent access to or use of shared data. They also stipulate the use of licensing terms for downstream innovations that would not prevent further research.

In the past decade, the controlled-access data access policy model has been used to share data from thousands of genomic research studies around the world. As of October 2018, NIH had provided over 5600 investigators access to 1025 studies, resulting in over 2460 peer-reviewed publications contributing significant advances to a wide range of fields (https://osp.od.nih.gov/scientific-sharing/facts-figures/). It has also been used to share and extract considerable research value from large collaborative international data resource projects, such as the International Cancer Genome Consortium [44] and the International Human Epigenome Consortium [45].

A constant challenge with controlled-access policies is to achieve fair and transparent access procedures. Several steps in the application procedure are more open to interpretation and have been the subject of debate and scrutiny [46,47]. These are the assessment of a researcher's ability or qualification to undertake the proposed analysis, with different levels of research training deemed necessary by different data access committees and offices, for example, whether access should be allowed to postdoctoral research trainees. The second step that has been controversial and is not universally condoned for access review is the assessment of the scientific merit of projects. This step derives from access procedures that were set up to share biobank resources that involve not only data but also limited resources such as biospecimens. Making the most scientifically valuable use of such materials is therefore important to fully respect the contribution of biobank participants. There still being inherent risk in providing controlled-access data for research, this review of a study's scientific merit can nevertheless be justified on the basis of providing for accurate risk versus benefit assessment by assessing the likely scientific advance from the research. However, this step, in particular, may exacerbate potential conflicts of interest in the review process. Many DACs and DAOs directly involve the researchers who produced the data in the review process and it could be their scientific competitors who are requesting access. Finally, sharing only with applicants at "recognized scientific institutions" can also raise access issues. Generally speaking, this is taken to include research institutions both public and private, so long as data have been consented and authorized for commercial or for-profit uses, but would exclude, for example, researchers working in industries the public is averse to sharing with (e.g., insurance, advertising). Centralized access mechanisms, such as those for dbGaP through the NIH, facilitate fair and consistent access review procedures and decisions.

It is worth noting that there have been few reported incidents of noncompliance with the controlled-access model in a decade of existence [36]. One incident involved a publication retraction due to a breach of publication moratorium [48]. While also used in open-access sharing, publication moratoria are more formal in the controlled-access policy model as they are often included in the data access agreements that investigators and their institutions agree to as a condition of access. Even though publication moratoria were adopted by the community as an incentive to encourage more data sharing, ultimately, they led to a situation in which the publication of scientific findings was unauthorized. The NIH has withdrawn its support for publication moratoria in its most recent updates to its Genomic Data Sharing Policy [13].

A major shortcoming of controlled access is that it is rather resource intensive, both on the data management side and from the point of view of applicants [49]. Applications typically take several months to be prepared by applicants, reviewed by data access committees and offices, and finalized with the execution of the data access agreement by the applicant's host organization. Three steps are particularly labor intensive. First, for the experts involved in the review of access applications, usually as volunteers, the review of individual research proposals is time consuming and often duplicative of prior or concomitant scientific and ethical reviews carried out by funding agencies and research ethics boards. An important aspect of these reviews nonetheless is to ensure the compatibility of the research with the stated consent of the research participants. Second, reviewing the credentials of scientists applying for every application is also redundant, especially when data access procedures are standardized and centralized as they are for dbGaP. Third, on the side of the applicants, the review and signing of the data access agreements between their institutions and the organization providing the shared data is often a bottleneck in the application process. The agreements frequently lead to negotiations regarding their content (e.g., clauses relating to liability) before their acceptance by research institutions, even though they are generally intended to be accepted as such. Finally, the controlled-access model also tends to "silo" data from individual research projects, as it is intended for access to one dataset for one study.

If over the past decade the controlled-access policy model has been the standard model for sharing data from most large-scale human genomic studies, there have been two notable exceptions: the 1000 Genomes Project and the Personal Genome Project, both of which opted for the open-access data-sharing model. The 1000 Genomes Project Consortium, followed by the International Genome Sample Resource, is providing genomic data from several world population groups as reference data on genomic variation in these populations [50–52]. The data do not have any association with the actual health of individuals and potential reidentification would rely on prior knowledge of a participant's genetic information, therefore the risk to participants, if it were reidentified, has been considered low. The Personal Genome Project is pioneering public involvement in, and access to, personalized genomic research. Participants contribute samples for DNA analysis and other health information for the purposes of producing an open-access resource of genomic and health-related data [53–55]. The data shared publicly presents a greater risk than 1000 Genomes Project data and the project hinges on a robust and extensive informed consent process. Participants are aware of the risks of potential reidentification and misuse of data and accept them.

4. Registered-access policy model

With the steady growth in big data and personalized medicine approaches to research, not only were controlled-access mechanisms increasingly under pressure due to the volume of applications, but researchers and funders began to question whether this data access model would continue to scale and enable the most widespread use of data in the long term. Furthermore, the pursuit of ever-larger datasets to analyze, along with developments in IT cloud storage and data processing pipelines, pointed to a data access model in which controlled datasets could be merged and analyzed together, and which could enable more "hypothesis-free" big data analytics.

Indeed, if we consider that there are opportunity costs to any reduction in either the number of researchers accessing data or the amount of research they perform using that data, a strong argument can be made that such restrictions although necessary must be proportionate. Was the controlled-access policy model stifling, or even simply slowing down, the progress of research? The new problematic created by technological and research developments also meant that data access would need to more readily facilitate data integration from different studies, and support research strategies such as machine learning. This led the community to re-examine controlled access with a view to assessing its strengths and limitations, and ultimately, determine whether an alternative data access policy model could be proposed.

A new data access policy model was therefore proposed to overcome these challenges and improve and streamline access to controlled data: the registered-access model. Consensus around this new model was built through extensive consultation and interdisciplinary teamwork within the Global Alliance for Genomics and Health (GA4GH) [56], which is an international coalition dedicated to improving human health by maximizing the potential of genomic medicine through effective and responsible data sharing, as founded on the *Framework for Responsible Sharing of Genomic and Health-Related Data* [57]. GA4GH regroups scientific institutions from across genomics as well as related industry (healthcare and IT) and patient advocacy organizations.

There are several important differences between the controlled-access and the registered-access policy models (see Fig. 2.2). First and foremost, registered access is based on the registration of researchers to obtain a "bona fide researcher" ("registered") status [43,58]. This would be the equivalent of providing qualified researchers with a "data passport" to gain access to data shared in this way [59], without reviewing their identity and professional activity for each data access.

Second, in the registered-access model, the data access agreements that are used in controlled access are replaced by acceptance of a set of standard good data use practices by researchers when registering. This is called the *Attestation* and encompasses the following statements:

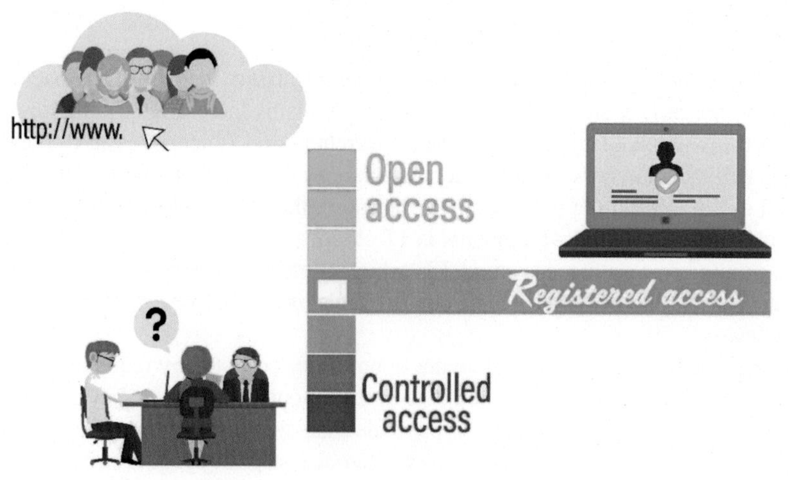

FIGURE 2.2

Three levels of data access policy: 1—open access, 2—registered access, and 3—controlled access. In open access, data are publicly and freely available from the internet. In registered access, only registered users have access to the data for uses within their "role" (researcher, clinical-care professional). Controlled access involves a review of applicants' individual research proposals, and applications are approved on a per-project basis.

1. "I am a bona fide researcher involved in biological/health research and will use these data for those purposes only."
2. "In my use of the data I will comply with all ethical and legal regulatory requirements applicable in my institution and country/region."
3. "My use of the data will be consistent with the GA4GH *Framework for Responsible Sharing of Genomic and Health-Related Data* (https://genomicsandhealth.org)."
4. "I will only use the data for the purposes allowed by the provider and I will abide by any consent conditions expressed as Consent Codes."
5. "I agree to forego any attempt to identify individuals represented in the dataset, except by prior written permission from the provider's sponsoring institution."
6. "I will treat the data as confidential and I will not share it with others."
7. "I will keep the data secure and delete all copies of the data when I no longer use the data or the permission period has expired."

There are plans to supplement these basic good data use practices with guidance, for example, for the security standards required to "keep the data secure" beyond deleting it after its use.

This registration process, along with the security of registered-access identities and claims, aims to achieve a high level of accountability from registered users. GA4GH is currently producing a standard technical specification for "bona fide researcher" access claims for authorization of access to registered data resources (GA4GH Passports) [59a].

Another important difference between the controlled-access policy model is that registered access is a role-based access model. Registered researchers are granted access to use data within all of their research activities and potential projects. In contrast to the controlled-access policy model, there is no review of proposed research uses for access. This change places much greater trust in researchers, and also in general research oversight mechanisms. Indeed, research activities are monitored at various stages of the research process and do not necessarily need to be checked again for the reuse of shared data. Importantly, it entails that researchers respect consent-related permissions in their use of the data (see discussion of Consent Codes later).

Several GA4GH "Driver" projects are using, or plan to use, registered access to facilitate access to data. ELIXIR (Europe's infrastructure for life science information) is piloting the model for access to research data as well as to high-performance computing resources [60]. Through its Beacon Project, which enables the discovery of genetic variants across multiple worldwide datasets, registered access will enable the sharing of more details than the simple existence of genetic variants (e.g., that they are present in individuals with a specific health condition) [61].

The Matchmaker Exchange (MME) is a federated network connecting databases of genomic and phenotypic data using a common application programming interface to facilitate rare disease gene discovery [62]. Registered access may help MME to expand its functionality and facilitate access to details of a matched case, including variants in a specific gene and phenotypic information.

The BRCA Challenge provides an umbrella under which many groups can collaborate and bring together data to improve the precision of assessing variants across both BRCA1 and BRCA2 [34]. Although its main resource on BRCA variant interpretation is publicly available, registered access could allow more sharing of the patient data supporting clinical interpretations with expert review teams, researchers, or clinicians.

Registered access is based on the premise that overall, broad access to research data is likely to bring about the greatest public benefits from research. This includes access to data by groups other than professional researchers, such as clinical-care professionals (e.g., working in clinical diagnostic laboratories), who rely on access to scientific evidence in their professional activities. Other groups such as patient advocacy groups/charities, scientific regulators, and reporters may also benefit from similar access arrangements, as could research participants and patients themselves along with other members of the public such as citizen scientists. However, adapting the registered access accountability model for a wide range of user "roles" presents considerable challenges and remains a long-term goal.

5. Ongoing concerns and developments

5.1 Maintaining consent: Consent Codes

While the models presented here attempt to regulate research data access, with the ever-greater sharing and reuse of research data, maintaining respect for the original consent provisions for use of participant data has become a critical issue. There are two important aspects to maintaining consent provisions with shared data: 1—consent-based permissions and restrictions must be specified with the shared data, and 2—consent provisions need to be clear to researchers and other potential users accessing the data to be respected.

In the controlled-access model, the review of individual research proposals allows data guardians to check that planned secondary uses of data respect participant consent. However, researchers accessing data through registered access must also respect consent, and facilitating this is a question of ongoing concern. Many of the current genomic projects that are generating data for use by the research community (i.e., projects with broad data-sharing policies) have recruited individuals who were willing to consent to very broad use of the data (e.g., use for all health-related or biomedical research purposes). However, access to older collections of data or to projects with a narrower consent scope rapidly extends the challenges to facilitate their appropriate reuse.

As an answer to these concerns, a list of Consent Codes has been developed by an international team working with GA4GH. The Consent Codes aim to address the recording of data access and use conditions stemming from study consent processes, their communication, and their understanding by all researchers [63]. Largely based on an analysis and categorization of existing data access conditions for the hundreds of controlled-access datasets archived by dbGaP and EGA, the Consent Codes, if successfully adopted and used, could facilitate tasks such as searching databases by consent permissions to select appropriate datasets for research studies. Several efforts are also currently attempting to facilitate the "machine-readability" of access conditions (GA4GH Data Use Ontology and Automatable Discovery and Access Matrix) [64] and Consent Codes could be a possible entry point. The main concern with such systems being that programs automating functions, such as "search data by consent permission," represent and interpret consent-based permissions and restrictions correctly.

5.2 Data-sharing risk assessment: choosing the right access level

As the value and benefits of sharing data grow with every additional dataset and reuse of data, so too does the risk to individuals. Prohibitions on obtaining an individual's genetic data without consent, and genetic (and other) nondiscrimination laws, provide important protection and can have a significant positive impact on the overall risk of data sharing for participants. They are nevertheless only effective at the national or regional level. Thus the protections afforded through data access policy along with efforts to assess and reduce the potential risk of reidentification and misuse of genomic research data remain very important.

In considering data access policy models and data to be shared, two aspects are critical: both reducing the risk of data reidentification (see next chapters—reidentification and security) as well as its potential to cause harm through misuse. An overarching goal is to assign a level of protection to data that is commensurate with this combined risk. For reducing the risk of potential harm from data sharing, determining the relative sensitivity of health information is perhaps the most pressing line of research. For example, current policy developments call for open access to aggregate and variant data associated with disease, except in the case of particularly stigmatizing conditions or vulnerable populations [65]. This raises the question of determining how the community could decide what constitutes particularly stigmatizing, or otherwise more or less sensitive health information, to achieve optimal protection and access. The concept of a Data Sharing Privacy Test to assign health-related data to categories of "sensitivity" is one potential solution to this challenge [66].

Today, data-sharing policy has come to play a central role in many areas of science and it continues to evolve in response to the scientific communities' ambitions. It is but one—albeit very important—part of broader Open Science policies that are emerging to open up many aspects of the scientific endeavor through greater sharing of research methods and resources (including not only data, but also software, materials, lab-books, etc.), and open publication and commercialization practices [67]. The open-access, controlled-access, and registered-access data access policy models discussed here, each with its strengths and limitations, are the basis upon which widespread access to scientific data currently rests.

References

[1] International Human Genome Sequencing Consortium, Lander ES, Linton LM, et al. Initial sequencing and analysis of the human genome. Nature 2001;409:860.
[2] Bentley DR. Genomic sequence information should be released immediately and freely in the public domain. Science 1996;274:533–4.
[3] Cook-Deegan R, Heaney C. Patents in genomics and human genetics. Annual Review of Genomics and Human Genetics 2010;11:383–425.
[4] HUGO. Summary of principles agreed at the first international strategy meeting on human genome sequencing. Bermuda. 1996.
[5] The Fort Lauderdale Report. Sharing data from large-scale biological research projects: a system of tripartite responsibility. 2003.
[6] Mouse Genome Sequencing Consortium, Waterston RH, Lindblad-Toh K, et al. Initial sequencing and comparative analysis of the mouse genome. Nature 2002; 420:520–62.
[7] International HapMap Consortium. The international HapMap project. Nature 2003; 426:789–96.
[8] Caulfield T, McGuire AL, Cho M, et al. Research ethics recommendations for whole-genome research: consensus statement. PLoS Biology 2008;6:e73.
[9] Toronto International Data Release Workshop Authors, Birney E, Hudson TJ, et al. Prepublication data sharing. Nature 2009;461:168–70.
[10] Dyke SO, Hubbard TJ. Developing and implementing an institute-wide data sharing policy. Genome Medicine 2011;3:60.
[11] Kleppner D, Sharp PA. Research data in the digital age. Science 2009;325:368.
[12] Boulton G, Rawlins M, Vallance P, Walport M. Science as a public enterprise: the case for open data. Lancet 2011;377:1633–5.
[13] NIH. In: NIH, editor. Genomic data sharing policy; 2014.
[14] Wellcome Trust: Policy on data, software and materials management and sharing. 2017.
[15] Genome Canada data release and resource sharing policy. 2008.
[16] Research Councils UK. Common principles on data policy. 2011.
[17] European Commission. EC Commission Recommendation of 17.7.2012 on access to and preservation of scientific information. 2012.
[18] Hrynaszkiewicz I, Norton ML, Vickers AJ, Altman DG. Preparing raw clinical data for publication: guidance for journal editors, authors, and peer reviewers. BMJ 2010;340: c181.
[19] Hanson B, Sugden A, Alberts B. Making data maximally available. Science 2011;331: 649.
[20] Standard cooperating procedures. Nature Genetics 2011;43:501.
[21] Hardwicke TE, Mathur MB, MacDonald K, et al. Data availability, reusability, and analytic reproducibility: evaluating the impact of a mandatory open data policy at the journal cognition. Royal Society Open Science 2018;5:180448.
[22] Canadian institutes of health research open access policy. 2013.
[23] Tennant J, Waldner F, Jacques D, Masuzzo P, Collister L, Hartgerink C. The academic, economic and societal impacts of Open Access: an evidence-based review. version 3; referees: 4 approved, 1 approved with reservations F1000Research 2016;5.
[24] Plan S; 2018. https://www.coalition-s.org/.
[25] Leinonen R, Sugawara H, Shumway M, International Nucleotide Sequence Database Consortium. The sequence read archive. Nucleic Acids Research 2011;39:D19–21.

[26] Oliver JM, Slashinski MJ, Wang T, Kelly PA, Hilsenbeck SG, McGuire AL. Balancing the risks and benefits of genomic data sharing: genome research participants' perspectives. Public Health Genomics 2012;15:106–14.
[27] Trinidad SB, Fullerton SM, Bares JM, Jarvik GP, Larson EB, Burke W. Genomic research and wide data sharing: views of prospective participants. Genetics in Medicine 2010;12:486–95.
[28] Privacy act (R.S.C., 1985, c. P-21), [privacy act], in: Canada (ed); 1985.
[29] Personal information protection and electronic documents act (R.S.C. 2000, c. 5), [PIPEDA], in: Canada (ed); 2000.
[30] Regulation (EU) 2016/679 of the European Parliament and of the Council of 27 April 2016 on the protection of natural persons with regard to the processing of personal data and on the free movement of such data, and repealing Directive 95/46/EC (General Data Protection Regulation). 2016.
[31] United States Genetic information nondiscrimination act. 2008.
[32] Canada Genetic non-discrimination act. 2016.
[33] Lek M, Karczewski KJ, Minikel EV, et al. Analysis of protein-coding genetic variation in 60,706 humans. Nature 2016;536:285–91.
[34] Cline MS, Liao RG, Parsons MT, et al. BRCA Challenge: BRCA Exchange as a global resource for variants in BRCA1 and BRCA2. PLoS Genetics 2018;14:e1007752.
[35] Bobrow M. Balancing privacy with public benefit. Nature 2013;500:123.
[36] Ramos EM, Din-Lovinescu C, Bookman EB, et al. A mechanism for controlled access to GWAS data: experience of the GAIN Data Access Committee. The American Journal of Human Genetics 2013;92:479–88.
[37] Milius D, Dove ES, Chalmers D, et al. The International Cancer Genome Consortium's evolving data-protection policies. Nature Biotechnology 2014;32:519–23.
[38] Mailman MD, Feolo M, Jin Y, et al. The NCBI dbGaP database of genotypes and phenotypes. Nature Genetics 2007;39:1181–6.
[39] Lappalainen I, Almeida-King J, Kumanduri V, et al. The European Genome-phenome Archive of human data consented for biomedical research. Nature Genetics 2015;47: 692–5.
[40] Kaye J, Heeney C, Hawkins N, de Vries J, Boddington P. Data sharing in genomics–re-shaping scientific practice. Nature Reviews Genetics 2009;10:331–5.
[41] Shabani M, Knoppers BM, Borry P. From the principles of genomic data sharing to the practices of data access committees. EMBO Molecular Medicine 2015;7(5):507–9.
[42] Tryka KA, Hao L, Sturcke A, et al. NCBI's database of genotypes and phenotypes: dbGaP. Nucleic Acids Research 2014;42:D975–9.
[43] Dyke SO, Kirby E, Shabani M, Thorogood A, Kato K, Knoppers BM. Registered access: a 'Triple-A' approach. European Journal of Human GeneticsJHG 2016;24:1676–80.
[44] International Cancer Genome Consortium, Hudson TJ, Anderson W, et al. International network of cancer genome projects. Nature 2010;464:993–8.
[45] Stunnenberg HG, International Human Epigenome Consortium, Hirst M. The international human Epigenome Consortium: a blueprint for scientific collaboration and discovery. Cell 2016;167:1897.
[46] Shabani M, Dyke SO, Joly Y, Borry P. Controlled access under review: improving the governance of genomic data access. PLoS Biology 2015;13:e1002339.
[47] Dyke SOM, Saulnier KM, Pastinen T, Bourque G, Joly Y. Evolving data access policy: the Canadian context. FACETS 2016;1:138.
[48] Holden C. Scientific publishing. Paper retracted following genome data breach. Science 2009;325:1486–7.

[49] Joly Y, Dove ES, Knoppers BM, Bobrow M, Chalmers D. Data sharing in the post-genomic world: the experience of the international cancer genome Consortium (ICGC) data access compliance office (DACO). PLoS Computational Biology 2012; 8:e1002549.
[50] Genomes Project Consortium, Abecasis GR, Altshuler D, et al. A map of human genome variation from population-scale sequencing. Nature 2010;467:1061–73.
[51] Clarke L, Zheng-Bradley X, Smith R, et al. The 1000 Genomes Project: data management and community access. Nature Methods 2012;9:459–62.
[52] Clarke L, Fairley S, Zheng-Bradley X, et al. The international Genome sample resource (IGSR): a worldwide collection of genome variation incorporating the 1000 Genomes Project data. Nucleic Acids Research 2017;45:D854–9.
[53] Church GM. The personal genome project. Molecular Systems Biology 2005;1. 2005 0030.
[54] Ball MP, Bobe JR, Chou MF, et al. Harvard Personal Genome Project: lessons from participatory public research. Genome Medicine 2014;6:10.
[55] Reuter MS, Walker S, Thiruvahindrapuram B, et al. The Personal Genome Project Canada: findings from whole genome sequences of the inaugural 56 participants. Canadian Medical Association Journal = journal de l'Association medicale canadienne 2018;190:E126–36.
[56] Global Alliance for Genomics and Health. GENOMICS. A federated ecosystem for sharing genomic, clinical data. Science 2016;352:1278–80.
[57] Knoppers BM. Framework for responsible sharing of genomic and health-related data. The HUGO Journal 2014;8.
[58] Dyke SOM, Linden M, Lappalainen I, et al. Registered access: authorizing data access. European Journal of Human GeneticsJHG 2018;26:1721–31.
[59] Cabili MN, Carey K, Dyke SOM, et al. Simplifying research access to genomics and health data with Library Cards. Scientific Data 2018;5:180039.
[59a] GA4GH Passports technical specification (v1.0.0), https://github.com/ga4gh-duri/ga4gh-duri.github.io/blob/master/researcher_ids/ga4gh_passport_v1.md.
[60] Linden M, Prochazka M, Lappalainen I, et al. Common ELIXIR service for researcher authentication and authorisation [version 1; referees: 3 approved, 1 approved with reservations] F1000Research 2018;7.
[61] Fiume M, Cupak M, Keenan S, et al. Federated discovery and sharing of genomic data using Beacons. Nature Biotechnology 2019;37:220–4.
[62] Philippakis AA, Azzariti DR, Beltran S, et al. The matchmaker Exchange: a platform for rare disease gene discovery. Human Mutation 2015;36:915–21.
[63] Dyke SO, Philippakis AA, Rambla De Argila J, et al. Consent Codes: upholding standard data use conditions. PLoS Genetics 2016;12:e1005772.
[64] Woolley JP, Kirby E, Leslie J, et al. Responsible sharing of biomedical data and biospecimens via the "automatable discovery and access matrix" (ADA-M). NPJ Genomic Medicine 2018;3:17.
[65] Health UNIo, editor. Update to NIH management of genomic summary results access NOT-OD-19-023; 2018.
[66] Dyke SOM, Dove ES, Knoppers BM. Sharing health-related data: a privacy test? NPJ Genomic Medicine 2016;1:16024.
[67] Poupon V, Seyller A, Rouleau GA. The tanenbaum open science institute: leading a paradigm shift at the montreal neurological institute. Neuron 2017;95:1002–6.

CHAPTER

Information leaks in genomic data: inference attacks

3

Erman Ayday, PhD

Case Western Reserve University, Department of Electrical Engineering and Computer Science, Cleveland, OH, United States; Bilkent University, Ankara, Turkey

1. Inference attacks on statistical genomic databases

One of the uses of genomic data is in research settings. Many researchers analyze large genomic datasets to discover associations between genes and known traits. Such studies are also known as genome-wide association studies (GWAS). The results of such case-control studies are typically shared via research papers. These shared results may include the IDs of strongly associated single nucleotide polymorphisms (SNPs) with a trait and the value of their associations, minor allele frequencies (MAFs) of the SNPs in the dataset, or correlations between the SNPs in the dataset. As these results are computed over a large number of individuals, public sharing of GWAS results was believed to be safe in terms of the privacy of study participants. However, in 2018, Homer et al. showed that if an adversary has access to the genomic data of a target, using the results of the GWAS study, the adversary can infer if the target individual was in the case group of the corresponding study [2]. This information also reveals the trait of the target individual that is associated with the case group. The adversary compares the target's SNPs with the reported MAFs of the case group and MAFs of a reference population that does not include the target. Then, it runs a statistical hypothesis test to determine the likelihood of whether the target is in the case group or not. The impact of this attack was significant such that as a result of this identified attack, NIH decided to remove all public aggregate genomic data from its website and started asking the researchers to sign a user agreement to access such aggregate data.

The attack identified by Homer et al. required around 10,000 SNPs (10,000 SNPs of the victim and MAF value of 10,000 SNPs from the case group of the study), which can be considered as unrealistic. However, later Wang et al. developed a more powerful attack that required much less SNP (around 200) [3] by utilizing the correlations between the SNPs in the genome (i.e., linkage disequilibrium). The common point of these attacks is that they use statistical information that is computed over the participants of genomic databases. Thus, as a countermeasure to such attacks, researchers proposed adding controlled noise to the statistical results

Responsible Genomic Data Sharing. https://doi.org/10.1016/B978-0-12-816197-5.00003-6

before publicly sharing them [4,5]. However, later, it has been shown that even much simpler queries to genomic databases create vulnerabilities for the database participants. Next, we discuss such vulnerabilities.

2. Inference attacks on genomic data-sharing beacons

The human genome is the utmost personal identifier and sharing genomic data for research while preserving the privacy of the individuals have been challenging many different fields (e.g., medicine, bioinformatics, computer science, law, and ethics) for long, due to possibly dire ethical, monetary, and legal consequences. To address this challenge and create frameworks and standards to enable the responsible, voluntary, and secure sharing of genomic and health-related data, the Global Alliance for Genomics and Health (GA4GH) was formed by the community [6]. The current genomic data-sharing standard that is implemented by GA4GH and being vastly used by the community is realized via genomic data-sharing beacons. Beacons are the gateways that let users (researchers) and data owners exchange information without—in theory—disclosing any personal information. A user who wants to apply for access to a dataset can learn whether individuals with specific alleles of interest are present in the beacon through an online interface. That is, the user submits a query, asking whether a genome exists in the beacon with a certain nucleotide at a certain position, and the beacon answers as "yes" or "no." If the dataset does not contain the desired variant, genomic data are not shared and distributed unnecessarily. In addition, the user does not have to go through the paperwork to obtain a dataset that will not be helpful. The GA4GH provides a shared beacon interface [7] that as of December 2018 provides access to 67 beacons and acts as a hub where researchers and data owners meet.

Beacons are typically associated with a particular phenotype (e.g., beacon of individuals that are HIV+ or breast cancer). Therefore, the presence of an individual in a particular beacon is considered as privacy-sensitive information and the main aim of the beacons is to protect this information. An attacker, using the responses of a beacon and genomic data of a victim, may try to infer the membership of the victim in a particular beacon by running a membership inference attack. Beacon framework sets a barrier against membership inference attacks for several reasons. Data access is highly restricted as the system allows only presence/absence queries for variants that make the system resistant to allele frequency-based attacks [3]. Moreover, the framework does not tie any response to any specific individual. Given possibly a large number of individuals in a beacon, a "yes" answer may originate from any one of the participants. Despite all these factors, several works have proven that beacons are not bulletproof and leak membership information.

In 2015, Shringarpure and Bustamante introduced a likelihood-ratio test (LRT) that can predict whether an individual is in the beacon, by repeatedly querying

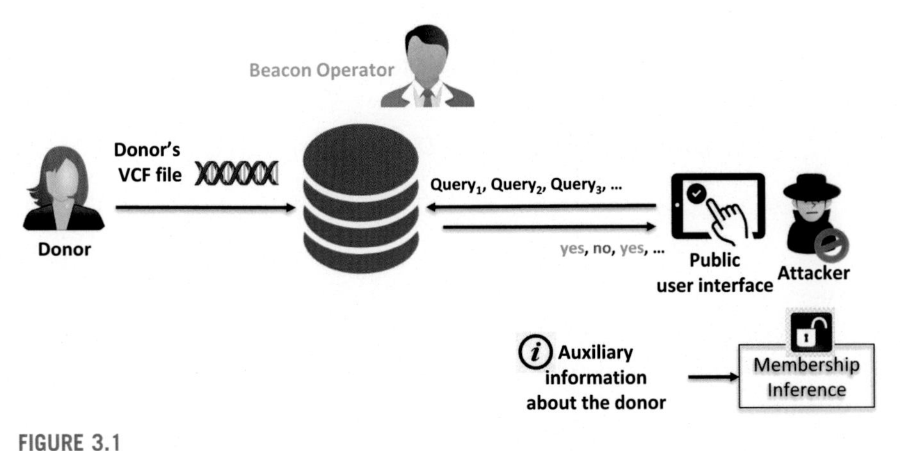

FIGURE 3.1

Membership inference attacks on genomic data-sharing beacons.

the beacon for 250 SNPs of a victim (SB attack) [8]. Note that inferring the membership of an individual in a beacon that is associated with a sensitive phenotype is equivalent to uncovering the sensitive phenotype about the victim. Then, Raisaro et al. showed that if the attacker also uses MAFs of the SNPs, they need fewer queries to reidentify a person by using SNPs with low MAF (optimal attack) [9]. In a recent work, we showed that even if the attacker does not have victim's low-MAF SNPs, it is still possible to do the membership inference by exploiting the correlations in the genome [10]. Furthermore, we showed that beacon responses can also be inferred using such correlations. A general framework summarizing these attacks is shown in Fig. 3.1. In the following, we provide some insights about our recent work.

We identified two novel vulnerabilities and we have shown that the privacy risk is even more serious than what has been shown previously in Refs. [8,9]. We considered two scenarios. In the first scenario, query inference attack (QI-attack) uses pairwise SNP correlations (linkage disequilibrium—LD) to infer the answers of future queries from previously answered queries. In this model, the attacker uses the LD value of an SNP pair to calculate the correlation of two minor alleles at the corresponding loci. The correlation is equal to the probability of the two minor alleles occurring together. On this basis, the attacker constructs an SNP network that uses weighted, directed edges between SNPs in high LD. First, the attacker selects the SNPs to be queried. This step is identical to the optimal attack and leads to a set of candidate SNPs S to be queried, starting from the lowest MAF SNP_i. Second, if any nonqueried SNP_j in S is a neighbor of SNP_i in the SNP network, the attacker infers the result of the query and does not pose a query for SNP_j. Let the null hypothesis (H_0) refer to the case in which the queried genome is not in the beacon and alternative hypothesis (H_1) be the case in which the queried genome is a member of the beacon. In the following, we present the log-likelihood (L) under the null and alternative hypothesis, which also integrates the inference error.

$$L_{H_0}(R) = \sum_{i=1}^{n}\left(x_i \log\left(1 - D_N^i\right) + (1 - x_i)\log\left(D_N^i\right) \right.$$
$$\left. + \sum_{j=1}^{m} \gamma x_i \log\left(1 - D_N^j\right) + \gamma(1 - x_i)\log\left(D_N^j\right)\right) \tag{3.1}$$

$$L_{H_1}(R) = \sum_{i=1}^{n}\left(x_i \log\left(1 - \delta D_{N-1}^i\right) + (1 - x_i)\log\left(\delta D_{N-1}^i\right) \right.$$
$$\left. + \sum_{j=1}^{m} \gamma x_i \log\left(1 - \delta D_{N-1}^j\right) + \gamma(1 - x_i)\log\left(\delta D_{N-1}^j\right)\right), \tag{3.2}$$

where R is the response set, x_i is the answer of the beacon to the query at position i(1 for yes, 0 for no), and δ represents a possible sequencing error. In addition, n is the number of posed queries, m is the number of neighbors that can be inferred for each posed query x_i, and γ corresponds to the confidence of the inferred answer, obtained from the SNP network. Furthermore, D_N^i is the probability that none of the N individuals in the beacon has the queried allele at position i. D_{N-1}^i represents the probability of no individual except for the queried person having the queried allele at position i. The computations of D_{N-1} and D_N depend on the queried position i and change at each query as follows: $D_{N-1}^i = (1 - f_i)^{2N-2}$ and $D_N^i = (1 - f_i)^{2N}$, where f_i represents the MAF of the SNP at position i. The LRT statistic, Λ, is then determined as

$$\Lambda = \sum_{i=1}^{n}\left(\log\left(\frac{D_N^i}{\delta D_{N-1}^i}\right) + \log\left(\frac{\delta D_{N-1}^i\left(1 - D_N^i\right)}{D_N^i\left(1 - \delta D_{N-1}^i\right)}\right)x_i + \sum_{j=1}^{m} \log\left(\frac{D_N^j}{\delta D_{N-1}^j}\right) \right.$$
$$\left. + \log\left(\frac{\delta D_{N-1}^j\left(1 - D_N^j\right)}{D_N^j\left(1 - \delta D_{N-1}^j\right)}\right)\gamma x i \right). \tag{3.3}$$

By not querying the beacon for the answers that can be inferred with high confidence, this model requires less number of queries compared to the optimal attack, while achieving the same response set.

In the second scenario, the genome inference attack (GI-attack) considers the case where individuals publicly share their genomes by taking necessary precautions, such as hiding their sensitive SNP positions with MAFs $< t$. The GI-attack performs allele inference to recover hidden SNP positions and infers alleles at the victim's hidden loci. The attacker uses a high-order Markov chain to model SNP correlations as described by our prior work [11].

Based on the victim's genome sequence, the attacker calculates the likelihood of the victim having a heterozygous SNP at the chosen SNP position i as $P_k(SNP_i) = P(SNP_i|SNP_{i-1}, SNP_{i-2}, \ldots, SNP_{i-k})$, where k is the order of the Markov chain. To use a high-order Markov chain to infer hidden SNPs, genome sequences from public sources such as the 1000 Genomes Project or HapMap can be used to train the model. Such publicly available genome datasets are typically available with the population information about their anonymized participants. We use a dataset that is consistent with the victim's population to build our high-order model. Accordingly, [11] defines the kth-order model as

$$P_k(SNP_i) = \begin{cases} 0 & \text{if } F\left(SNP_{i-k,i-1}\right) = 0 \\ \dfrac{F\left(SNP_{i-k,i}\right)}{F\left(SNP_{i-k,i-1}\right)} & \text{if } F\left(SNP_{i-k,i-1}\right) > 0, \end{cases} \tag{3.4}$$

where $F\left(SNP_{i,j}\right)$ is the frequency of occurrence of the sequence that contains SNP_i to SNP_j. The SNPs are ordered according to their physical positions on the genome. The model works by comparing the SNPs in $SNP_{i-1,j}$ which are before SNP_i on the genome sequence to the same SNP positions in the training dataset. If the training set contains other genomes with the same SNP sequence and these sequences are followed by a heterozygous position, we can calculate the probability of SNP_i being heterozygous for the victim. If the calculated likelihood of the victim having a heterozygous position is high enough (in this case equal to 1), the attacker queries the beacon for the inferred SNP position, starting from the SNP with the lowest MAF.

To evaluate the identified vulnerabilities, we tested and compared our methods on (i) a simulated beacon, and (ii) the beacons of the beacon network [7] operated by GA4GH beacon network. We worked on a simulated beacon with 65 people from the Utah Residents (CEPH) with Northern and Western European Ancestry (CEU) population of the HapMap dataset. While testing for the alternative hypothesis, we used 20 randomly picked people from the beacon. For the null hypothesis, we used 40 additional people from the same population of the HapMap project. The LD scores, allele frequencies, and genotype data were also obtained from the CEU dataset of the HapMap project [12]. For the GI-attack, we used a fourth-order Markov chain as also done in Ref. [11]. We show the power curves for the optimal attack [8], the QI-attack, and the GI-attack each at 5% false-positive rate in Fig. 3.2. We observed that the QI-attack requires 30% less number of queries on average compared to the optimal attack. In addition, the GI-attack requires only five queries for all tested thresholds of t (larger t means less available information, which also means weaker attacker). We do not show the results of the SB attack [9] in the figure because we observed that the SB attack requires 1400 queries even when $t = 0$.

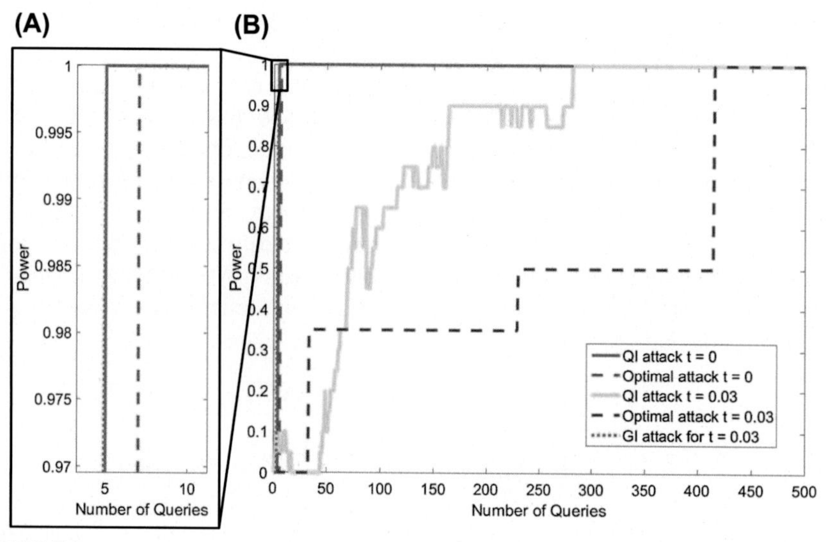

FIGURE 3.2

(A) Close-up of the power curves, where the number of queries is <10. (B) Power curves of the optimal attack [9], the QI-attack, and the GI-attack for different thresholds of *t* on a beacon with 65 members constructed with individuals from the CEU dataset of the HapMap project. *t* indicates the threshold up to which SNPs with an MAF $< t$ are hidden as a countermeasure [10].

3. Inference attacks on kin genomic privacy

The genome of an offspring is inherited from his/her parents. Thus, genomic data are correlated between family members. This correlation and other kinship-related properties of genomic data result in additional vulnerabilities for genomic data. For instance, using the fact that Y-chromosome is directly inherited among the male-line of a family, public genealogy databases, and other auxiliary information sources, Gymrek et al. showed that it is possible to deanonymize the participants of anonymized genomic datasets [13]. In the following, we detail two inferences on genomic privacy that utilizes the kinship relationship between the family members.

Kinship inference from public genomic databases: Another threat is the identification of kinship relationships between anonymous donors in public genomic databases. In a previous work, we defined two routes that leak kinship information from publicly available databases [14]. We showed how the kinship relationship between participants of anonymous genomic databases can be efficiently identified using (i) genotype similarity and (ii) outlier allele pair counts. We showed that the relatedness of two individuals can be inferred based on their genotype similarity using a kinship metric. We observed that such kinship metrics are mostly dominated by the number of SNPs that are heterozygous in both individuals. Thus, before publicly sharing data, positions wherein the two individuals are found to be heterozygous can be hidden as it decreases the kinship coefficient between two family members effectively. However, we also showed that this alone will cause another privacy leakage as the number of positions where the two family members are heterozygous will be too small. Simply comparing this number to the population, one could infer that the two individuals are indeed in the same family. Thus, we proposed a technique to protect kinship privacy against these risks while maximizing the utility of shared data. We developed an optimization-based framework for the sharing of information. The method involves systematic identification of minimal portions of genomic data to mask as new participants are added to the database. Choosing the proper positions to hide is cast as an optimization problem in which the number of positions to mask is minimized subject to privacy constraints that ensure the familial relationships are not revealed. The proposed scheme is also shown in Fig. 3.3, where g_i represents the genotype of an individual i, f_k represents the family that individual i belongs to, and g_i' represents the partially shared genotype of i (after selectively hiding some SNPs considering the privacy risks).

Genome inference using kinship relationships: Individuals (and their family members) share (partial) genomic data on public platforms. However, using special characteristics of genomic data, background knowledge that can be obtained from the web, and family relationships between the individuals, it is possible to infer the hidden parts of shared (and unshared) genomes. In earlier work, considering simple correlations in the genome (as well as Mendel's law and partial genomes of a victim and his family members), we showed how an adversary can efficiently and accurately infer a target's genome [15].

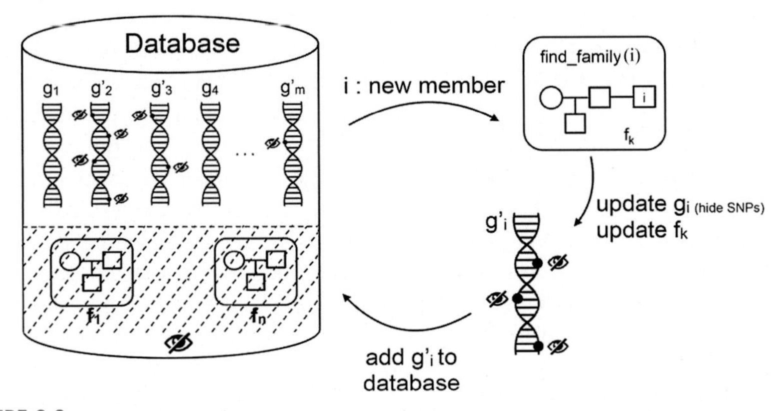

FIGURE 3.3

Overview of the proposed scheme in Ref. [14]. When a new person i with genotype g_i is added to the database, we check for is relatives in the database and determine the family i is related to. The privacy of the family f_k (to which i belongs to) is protected by selectively hiding a portion of g_i. The genotype of person i is then partially shared and this partially shared genotype is denoted as g'_i.

The goal of the adversary is to infer some *targeted SNPs* of a member (or multiple members) of a *targeted family*. Let $\mathbf{F}$ be the set of family members in the targeted family (whose family tree is $\mathcal{G}_{\mathbf{F}}$) and $\mathbf{S}$ be the set of SNP IDs (on the DNA sequence), where $|\mathbf{F}| = n$ and $|\mathbf{S}| = m$. Let also x_j^i be the value of SNP $j(j \in \mathbf{S})$ for individual $i(i \in \mathbf{F})$, where $x_j^i \in \{0, 1, 2\}$. In addition, $\mathbb{X}$ is an $n \times m$ matrix that stores the values of the SNPs of all family members. Among the SNPs in $\mathbb{X}$, the ones whose values are unknown are in set $\mathbb{X}_{\mathrm{U}}$, and the ones whose values are known (by the adversary) are in set $\mathbb{X}_{\mathrm{K}}$. $\mathcal{F}_R\left(x_j^M, x_j^F, x_j^C\right)$ is the function representing the Mendelian inheritance probabilities, where (M, F, C) represent mother, father, and child, respectively. Finally, $\mathbf{P} = \left\{p_j^b : i \in \mathrm{S}\right\}$ represents the set of minor allele probabilities (or MAF) of the SNPs in $\mathbf{S}$. The adversary carries out a reconstruction attack to infer $\mathbb{X}_{\mathrm{U}}$ by relying on his background knowledge, $\mathcal{F}_R\left(x_j^M, x_j^F, x_j^C\right), \mathbb{L}, \mathbf{P}$, and on his observation $\mathbb{X}_{\mathrm{K}}$. Here, $\mathbb{L}$ is a $m \times m$ matrix representing the pairwise LD between each pair of SNPs. We formulated this reconstruction attack as finding the marginal probability distributions of unknown variables $\mathbb{X}_{\mathrm{U}}$, and to run this attack in an efficient way, we formulated the problem on a factor graph and use the belief propagation algorithm for inference. Belief propagation [16] is a message-passing algorithm for performing inference on graphical models (e.g., Bayesian networks or Markov random fields). It is typically used to compute marginal distributions of unobserved variables conditioned on the observed ones.

In a later work, we further improved the existing work on inference attacks on kin genomic privacy [17]. We considered two different complex correlation models in the genome: (i) Markov chain, in which we consider the genome as a sequence of SNPs, where the value of each SNP depends on the values of neighboring k SNPs; (ii) hidden Markov model (HMM), in which the genome is modeled as a Markov process with unobserved (hidden) states. We realize the HMM model for the genome by using the recombination model [18]. We also utilized the phenotype information about the victims. We proposed an efficient message-passing algorithm to consider all aforementioned background information for the inference. We assume that the attacker has access to the following resources about the target individuals: (i) the partial genomic data of individuals (from public genomic databases and genome-sharing websites), (ii) phenotype information (physical characteristics) of individuals from online social networks (OSNs), (iii) health-related information of individuals from OSNs and health-related social networks, and (iv) family bonds of individuals (e.g., their family trees) from OSNs or genealogy websites. Our proposed framework is also sketched in Fig. 3.4. The objective is to infer the missing parts of the genomes of target individuals. For this, we use family bonds between the individuals in the target set, the probabilistic relationship between the phenotype and genotype, a similar relationship between diseases and the genotype, and some genomic tools for inference such as high-order correlations in the genome and the recombination model. To run this inference attack efficiently, similar to the previous

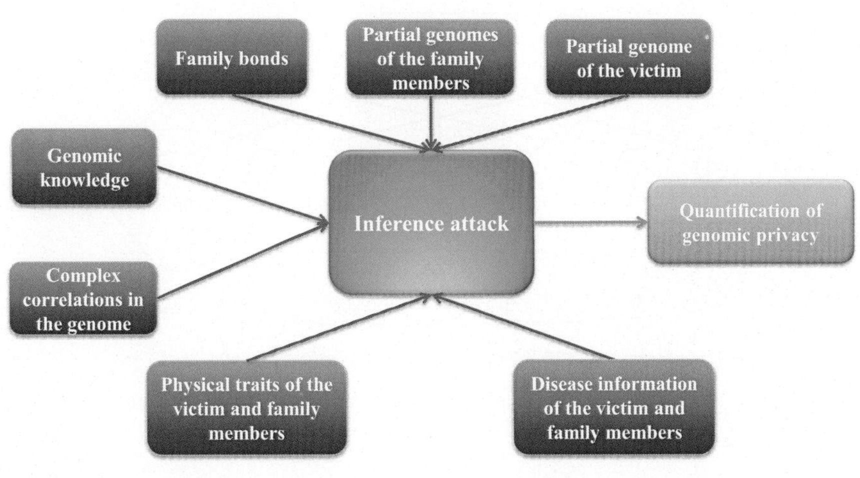

FIGURE 3.4

Overview of the proposed framework for the quantification of genomic privacy in Ref. [17].

work [15], we rely on the belief propagation algorithm on a factor graph. Then, we quantify the genomic privacy of individuals and show the risk for each individual.

A factor graph is a bipartite graph containing two sets of nodes (corresponding to variables and factors) and edges connecting these two sets. We form a factor graph by setting a variable node for each SNP x_j^i ($j \in \mathbf{S}$ and $i \in \mathbf{F}$). We use three types of factor nodes: (i) *familial factor node*, representing the familial relationships and reproduction, (ii) *correlation factor node*, representing the higher-order correlations between the SNPs by using either a Markov chain or hidden Markov model, and (iii) *phenotype factor node*, representing the correlation between the SNPs and the phenotypes (e.g., physical traits or diseases) of individuals. The factor graph representation of our proposed framework is shown in Fig. 3.5. We summarize the connections between the variable and factor nodes below:

- Each variable node x_j^i has its familial factor node f_j^i if at least one parent of individual i is in the target family. Furthermore, x_j^k $(k \neq i)$ is also connected to the familial factor node of x_j^i if k is the mother or father of i. If an individual i's both parents are not present in the target family, we do not assign familial factor nodes corresponding to the variable nodes of that individual. For example, in Fig. 3.5, all familial factor nodes belong to the child as his parents are present in the toy example. However, father's and mother's variable nodes do not have separate familial factor nodes.
- Variable nodes in set **C** are connected to a correlation factor node $g_{\mathbf{C}}^i$ (of individual i) if SNPs in **C** have correlation among each other. In particular, we consider higher-order correlations in the genome. We model these correlations either using a Markov chain or a hidden Markov model, HMM (i.e., recombination model). When we use a Markov chain with the order of k the correlation

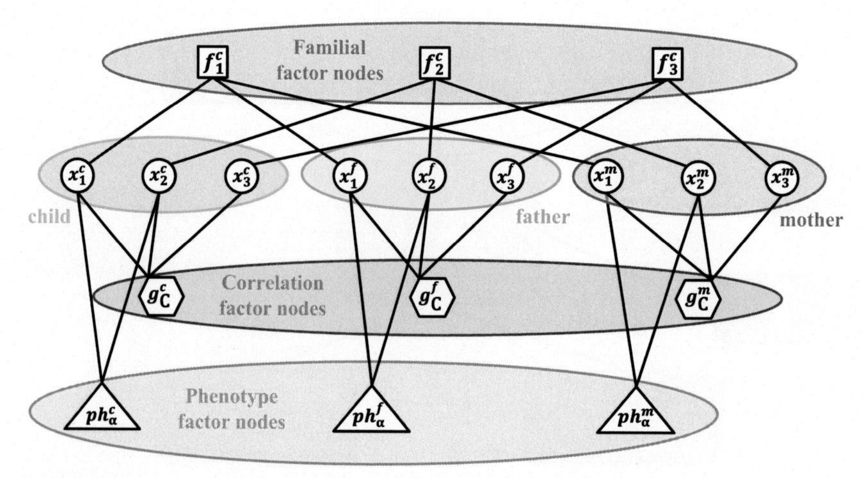

FIGURE 3.5

Factor graph representation of the proposed framework in [11].

set of node i equals $\mathbf{C}_i = \{node_{i-k}, node_{i-k+1}, node_{i-k+2}, \ldots, node_{i-1}\}$ if $i > k$, and $\mathbf{C}_i = \{node_1, node_2, node_3, \ldots, node_{i-1}\}$ if $i \leq k$, and when we use HMM, $\mathbf{C}$ includes all SNPs in a chromosome.

- Variable nodes of individual i in set $\mathbf{H}^i_\alpha$ connected to a phenotype factor node ph^i_α if SNPs in $\mathbf{H}^i_\alpha$ associated with the phenotype ph_α. Note that more than one SNP can be associated with a given phenotype. Similarly, an SNP may be associated with more than one phenotype.

As shown in Ref. [19], following the rules of belief propagation, the global probability distribution of the variable nodes can be factorized into products of local functions that are defined by the factor nodes following the rules of the belief propagation algorithm. The iterative belief propagation algorithm is based on exchanging messages between the variable and the factor nodes. We represent these messages as in the following:

- The message $\mu^{(v)}_{i\to k}\left(x^{i(v)}_j\right)$ (from a variable node i to a factor node k) denotes the probability of $x^{i(v)}_j = \ell(\ell \in \{0,1,2\})$, at the vth iteration.
- The message $\lambda^{(v)}_{k\to i}\left(x^{i(v)}_j\right)$ (from a familial factor node to a variable node) denotes the probability that $x^{i(v)}_j = \ell$, for $\ell \in \{0,1,2\}$, at the vth iteration given $\mathcal{F}_R\left(x^M_j, x^F_j, x^C_j\right)$, **P**, and the values of SNP j for the other two family members (other than individual i) that are connected to the corresponding familial factor node.
- The message $\beta^{(v)}_{k\to i}\left(\mathbf{C}, x^{i(v)}_j\right)$ (from a correlation factor node to a variable node) denotes the probability that $x^{i(v)}_j = \ell$, for $\ell \in \{0,1,2\}$, at the vth iteration given the high-order correlation between the SNPs in set $\mathbf{C}$.

- The message $\delta_{k\to i}^{(v)}\left(x_j^{i(v)}\right)$ (from a phenotype factor node to a variable node) denotes the probability that $x_j^{i(v)} = \ell$, for $\ell \in \{0,1,2\}$, at the vth iteration given the phenotype ph_k for individual i and the association of the corresponding phenotype with SNP j.

At each iteration of the algorithm, all variable and factor nodes generate their messages and send them to all of their neighbors as described earlier. At the end of each iteration, we compute the marginal probabilities of each variable node (by multiplying all incoming messages), and we stop the algorithm when the values of the marginal probabilities stop changing. Note that the computational complexity of this inference attack is linear with the number of variable or factor nodes in the factor graph. Overall, our results show that the attacker's inference power (on the genomic data of individuals) significantly improves by using complex correlations and phenotype information (along with information about their family bonds).

4. Inference attacks using genotype–phenotype associations

As discussed, genome-wide association studies aim to discover the associations between SNPs in the genome and other traits (e.g., physical attributes or diseases). Such genotype–phenotype associations also introduce a new kind of vulnerability for genomic data donors. Using such associations, publicly available (anonymized) genomic databases, and auxiliary information about the phenotypes of target individuals, an attacker can link the anonymous genome of a donor to his/her phenotype (and hence deanonymize the donor). Harmanci and Gerstein proposed frameworks to quantify the risk for such an attack using phenotype datasets [20]. They showed how an adversary can cross-reference genotype and phenotype datasets and deanonymize individuals accordingly. In our earlier work, we also tackled the same problem and quantified this privacy risk. In the following, we provide details about our work on this domain [21].

We proposed new deanonymization attacks that make use of only the most common piece of genomic information that is output today by major direct-to-consumer providers: the SNPs. The attack relies upon the fact that our SNPs are intrinsically linked to our phenotypic traits (such as eye color, blood type, or genetic diseases) and that genomic research progress provides us with more information about these links. For instance, the relationship between SNPs and phenotypes is increasingly used in forensics for reconstructing facial composites from DNA information [22,23]. If an adversary has access to phenotypic traits (e.g., visible traits) of an identified individual, he can use known correlations between phenotypic traits and genomic data to identify the genotype of this individual in a genomic database and to infer other sensitive information (such as predispositions to severe diseases) by using the deanonymized genomic data, as illustrated in Fig. 3.6. The adversary

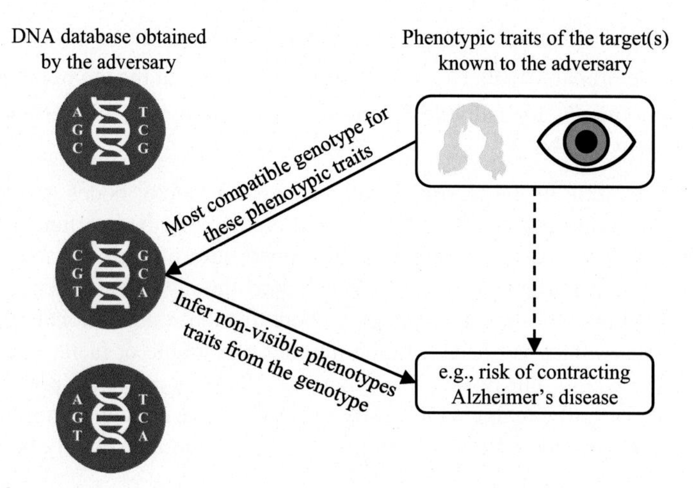

FIGURE 3.6

Illustration of the identification attack in Ref. [21]: The adversary identifies the genotypes of a target individual from some of her visible phenotypic traits and uses the deanonymized genotype to infer her susceptibility to Alzheimer's disease.

also has access to anonymized genotypes through a collaborative genome-sharing platform (such as the Personal Genome Project) and want to deanonymize them by relying upon phenotypic information gathered on OSNs. The matching between OSNs and genomic profiles is (even) easier if, for example, the ZIP code is available with the genomic profiles, thus enabling the OSN profiles to be filtered before the matching attack. The adversary might also want to match different online identities, for example, open SNP profiles that contain genomic data with PatientsLikeMe (https://www.patientslikeme.com) profiles that contain phenotypic information (mainly health condition, such as diseases).

We study two deanonymization attacks: (i) the identification attack, where the adversary wants to identify the genotype (among multiple genotypes) that corresponds to a given phenotype, and (ii) the perfect matching attack, where the adversary wants to match multiple phenotypes to their corresponding genotypes. We rely upon analytical tools for maximizing the matching likelihood in both attacks, and we assume two types of background knowledge: one that makes use of existing genetic knowledge from the association between SNPs and phenotypic traits (unsupervised approach), and another that learns the genomic–phenotypic statistical relationships from datasets containing both data types (i.e., genomic and phenotypic).

The adversary is assumed to have access to two distinct datasets: (i) a set $\mathcal{G} = \{\vec{g}_1, \vec{g}_2, \ldots, \vec{g}_n\}$ of genotypes of n different individuals, where $\vec{g}_i = (g_{i,1}, \ldots, g_{i,s})$ is a vector containing the SNP values of individual i, with $g_{i,j} = \{0, 1, 2\}$ and (ii) a set $\mathcal{P} = \{\vec{p}_1, \vec{p}_2, \ldots, \vec{p}_m\}$ of phenotypes of m individuals, where $\vec{p}_i = (p_{i,1}, \ldots, p_{i,t})$ is a vector containing the values of phenotypic traits of i. Note that $p_{i,j} \in \mathcal{P}_j$, where $\mathcal{P}_j$ is the set of values trait j can take. For instance, if trait j represents eye color, we could have $\mathcal{P}_j = \{\text{"}brown\text{"}, \text{"}blue\text{"}, \text{"}green\text{"}\}$ Not all participants have all their s SNP values known, or all t traits accessible. In the following, we provide the details of the proposed perfect matching attack.

In the perfect matching attack, the goal of the adversary is to assign precisely one genotype to one phenotype, such that the resulting n assignments maximize the product of the likelihoods $\prod_{i=1}^{n} \Pr[\vec{g}_i]\vec{p}_{\sigma(i)}$ over all $n!$ assignments σ between size-n sets $\mathcal{G}$ and $\mathcal{P}$. Hence, the assignment σ^* that maximizes likelihood is

$$\sigma^* = \underset{\sigma}{\operatorname{argmax}} \prod_{i=1}^{n} \prod_{l=1}^{t} \prod_{r \in \mathcal{R}_l} \Pr[g_{i,r}] p_{\sigma(i),l}. \tag{3.5}$$

Simply put, this problem is finding a perfect matching on a weighted bipartite graph, with n vertices on one side representing the n different genotypes, and n vertices on the other side representing the n phenotypes, as shown in Fig. 3.7. A weight is assigned to every edge of the complete bipartite graph. We define the weight $w_{i,j}$ between a genotype vertex G_i and a phenotype vertex P_j as the log-likelihood between genotype $\vec{g}_i$ and phenotype $\vec{p}_j$:

$$w_{i,j} := \log \Pr[\vec{g}_i]\vec{p}_j = \log \prod_{l=1}^{t} \prod_{r \in \mathcal{R}_l} \Pr[g_{i,r}] p_{j,l}, \tag{3.6}$$

and we solve the following optimization problem:

$$\sigma^* = \arg\max_{\sigma} \sum_{i=1}^{n} w_{i,\sigma(i)} \tag{3.7}$$

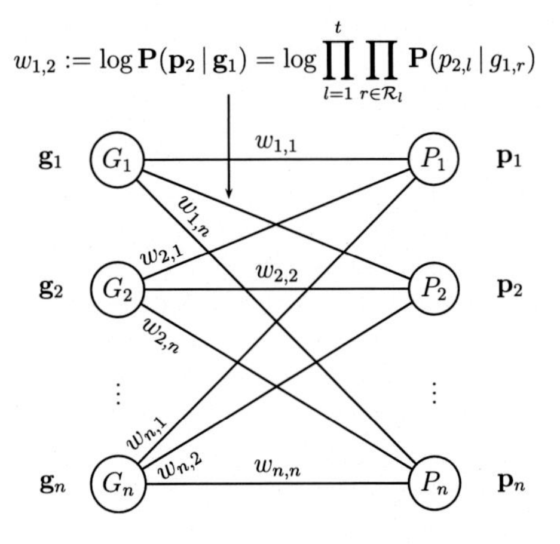

FIGURE 3.7

Complete weighted bipartite graph with n vertices on the left-hand side representing the individuals' genotypes and n other vertices on the right-hand side representing their phenotypes [21]. The weight $w_{i,j}$ is the log-likelihood of phenotype $\vec{p}_j$ given genotype $\vec{g}_i$.

$$= \arg\max_{\sigma} \sum_{i=1}^{n} \log \prod_{l=1}^{t} \prod_{r \in \mathcal{R}_l} \Pr[g_{i,r}] p_{\sigma(i),l} \tag{3.8}$$

$$= \arg\max_{\sigma} \log \prod_{i=1}^{n} \prod_{l=1}^{t} \prod_{r \in \mathcal{R}_l} \Pr[g_{i,r}] p_{\sigma(i),l} \tag{3.9}$$

The formulation in Eq. (3.9) enables us to maximize the sum of the weights instead of their product. Many existing algorithms can find the solution to this optimization problem in polynomial time. Here, we use the blossom algorithm that finds the maximum weight assignment in $O(n^3)$ [24]. We quantify the success of our attacks with different metrics. The proportion of pairs correctly matched reflects the correctness of the deanonymization attacks in general. In the identification attack target at individual j, the proportion of correct matches is equal to 1 if

$$j = \arg\max_{i} \Pr[\vec{g}_i] \vec{p}_j \tag{3.10}$$

and 0 otherwise. In the perfect matching attack, we measure the proportion of correct matches, that is, the ratio between the number of pairs correctly matched (i.e., where $\sigma^*(i) = i$), and the total number of matched pairs n.

Our experimental results show, with a database of 80 participants, in the identification attack, a proportion of 13% correct matches in the supervised case, and 5% in the unsupervised case. These results constitute a significant improvement: they outperform the considered baseline by a factor of eight and three, respectively. When the database size decreases to 10, the attack success increases to around 44% in the unsupervised case, and 52% in the supervised case. We also evaluate the adversary's ability to predict the predisposition to Alzheimer's disease of the database participants. With 10 participants, the average error of the adversary is halved when using the identification attack.

In the perfect matching attack, the proportion of correct matches is slightly better than in the identification attack: 16% in the supervised case and 8% in the unsupervised case. For a database of size 10, this proportion increases to 65% and 58%, respectively. With this size, the proportion of correct match is around four times higher than the baseline for both supervised and unsupervised approaches. We also evaluate the impact of the distinguishability between two individuals on the success of the perfect matching attack. Our results clearly show that the more distinguishable two individuals are, the more likely their genomic (or phenotypic) data will be deanonymized. This leads us to conclude that the threat to genomic privacy posed by our deanonymization attacks will become even more serious in the near future when more SNP-trait association information is discovered by genomic researchers, and available to the adversary.

5. Conclusions

In this chapter, we have discussed the existing inference attacks on genomic data [1]. As we have shown via several examples, public availability of genomic data raises serious privacy concerns. On the other hand, public availability of genomic data is very important to foster research. Thus, research on protection techniques against these privacy risks is crucial and existing protection techniques include (i) cryptographic solutions for private pattern-matching and the comparison of genomic sequences [25–28], (ii) privacy-preserving solutions to align and manage raw genomic data [29–31], (iii) privacy-preserving techniques to find similar patients in a distributed environment [32], (iv) private clinical genomics [33,34], and (v) privacy-preserving genomic research [4,35].

Unfortunately, such protection techniques do not eliminate the risk completely because inference attacks on genomic data are not limited to the ones that are discussed in this chapter. Although, to the best of our knowledge, we tried to highlight the prominent works on this field, such attacks are likely to evolve with (i) new discoveries in the field of genomics (e.g., new correlation models or new associations), (ii) new uses of genomic data (e.g., new direct-to-consumer genetic services), and (iii) new databases (e.g., evolving datasets that add or delete participants over time). Most existing protection techniques will most likely be insufficient as the attacks evolve, and hence new research will be required to develop new protection techniques.

Furthermore, new research is needed to convey the identified privacy risks and protection techniques to the end-user (e.g., genomic data donor) in a practical way. For instance, we have shown how data shared by family members threaten the genomic privacy of an individual. As a countermeasure against this risk, it may be argued that when donating genomic data, consent from (close) family members are required. However, the practicality of such a procedure is very questionable (especially considering unborn children or deceased relatives). It is also challenging for users to understand how to control what to share on social networks that can be used in attacks using genotype–phenotype associations. Thus, new and novel privacy risk quantification techniques are required to warn individuals against the risk of such attacks.

Acknowledgments

I would like to thank to my coauthors in the aforementioned works: Ercument Cicek, Iman Deznabi, Jean-Pierre Hubaux, Joachim Hugonot, Kevin Huguenin, Mathias Humbert, Nazanin Jafari, Mohammad Mobayen, Oznur Tastan, Amalio Telenti, and Nora von Thenen.

References

[1] Naveed M, Ayday E, Clayton EW, Fellay J, Gunter CA, Hubaux J-P, Malin BA, Wang X. Privacy in the genomic era. ACM Computing Surveys 2015;48(1).

[2] Homer N, Szelinger S, Redman M, Duggan D, Tembe W. Resolving individuals contributing trace amounts of DNA to highly complex mixtures using high-density SNP genotyping microarrays. PLoS Genetics 2008;4.

[3] Wang R, Li YF, Wang X, Tang H, Zhou X. Learning your identity and disease from research papers: information leaks in genome wide association study. In: CCS'09: Proc. of the 16th ACM conf. on computer and communications security; 2009. p. 534–44.
[4] Johnson A, Shmatikov V. Privacy-preserving data exploration in genome-wide association studies. Proceedings of KDD 2013;13:1079–87.
[5] Uhler C, Slavkovic A, Fienberg SE. Privacy-preserving data sharing for genome-wide association studies. Journal of Privacy and Confidentiality 2013;5(1).
[6] https://www.ga4gh.org/about-us/.
[7] http://beacon-network.org.
[8] Shringarpure SS, Bustamante CD. Privacy risks from genomic data-sharing beacons. The American Journal of Human Genetics 2015;97(5):631–46.
[9] Raisaro JL, Tramer F, Zhanglong J, Bu D, Zhao Y, Carey K, Lloyd D, Sofia H, Baker D, Flicek P, Shringarpure SS, Bustamante CD, Wang S, Jiang X, Ohno-Machado L, Tang H, Wang X, Hubaux J-P. Addressing beacon re-identification attacks: quantification and mitigation of privacy risks. The Journal of the American Medical Informatics Association 2016;24(4):799–805.
[10] von Thenen N, Ayday E, Cicek AE. Re-identification of individuals in genomic data-sharing beacons via allele inference. Bioinformatics 2019;35(3):365–71.
[11] Samani SS, Huang Z, Ayday E, Elliot M, Fellay J, Hubaux J-P, Kutalik Z. Quantifying genomic privacy via inference attack with high-order SNV correlations. In: Security and privacy workshops (SPW), 2015 IEEE; 2015. p. 32–40.
[12] Gibbs RA, Belmont JW, Hardenbol P, Willis TD, Yu F, Yang H, Ch'ang L-Y, Huang W, Liu B, Shen Y, et al. The international HapMap project. Nature 2003;426(6968): 789–96.
[13] Gymrek M, McGuire AL, Golan D, Halperin E, Erlich Y. Identifying personal genomes by surname inference. Science 2013;339(6117):321–4.
[14] Kale G, Ayday E, Tastan Ö. A utility maximizing and privacy preserving approach for protecting kinship in genomic databases. Bioinformatics 2018;34(2):181–9.
[15] Humbert M, Ayday E, Hubaux J-P, Telenti A. Addressing the concerns of the lacks family: quantification of kin genomic privacy. In: Proceedings of the 2013 ACM SIGSAC conference on Computer & communications security. ACM; 2013. p. 1141–52.
[16] Pearl J. Probabilistic reasoning in intelligent systems: networks of plausible inference. Morgan Kaufmann Publishers, Inc.; 1988.
[17] Deznabi I, Mobayen M, Jafari N, Tastan O, Ayday E. An inference attack on genomic data using kinship, complex correlations, and phenotype information. IEEE/ACM Transactions on Computational Biology and Bioinformatics (TCBB) 2018;15(4): 1333–43.
[18] Li N, Stephens M. Modeling linkage disequilibrium and identifying recombination hotspots using single-nucleotide polymorphism data. Genetics 2003;165.
[19] Kschischang F, Frey B, Loeliger HA. Factor graphs and the sum-product algorithm. IEEE Transactions on Information Theory 2001;47.
[20] Harmanci A, Gerstein M. Quantification of private information leakage from phenotype genotype data: linking attacks. Nature Methods 2016;13:251–6.
[21] Humbert M, Huguenin K, Hugonot J, Ayday E, Hubaux J-P. De-anonymizing genomic databases using phenotypic traits. PoPETs 2015;2015:99–114.
[22] Claes P, Hill H, Shriver MD. Toward DNA-based facial composites: preliminary results and validation. Forensic Science International: Genetics 2014;13:208–16.

[23] Claes P, Liberton DK, Daniels K, Rosana KM, Quillen EE, Pearson LN, McEvoy B, Bauchet M, Zaidi AA, Yao W, et al. Modeling 3D facial shape from DNA. PLoS Genetics 2014;10(3):e1004224.

[24] Galil Z. Efficient algorithms for finding maximum matching in graphs. ACM Computing Surveys (CSUR) 1986;18(1):23–38.

[25] Troncoso-Pastoriza JR, Katzenbeisser S, Celik M. Privacy preserving error resilient DNA searching through oblivious automata. Proceedings of ACM CCS 2007;07.

[26] Blanton M, Aliasgari M. Secure outsourcing of DNA searching via finite automata. In: DBSec'10: Proceedings of the 24th annual IFIP WG 11.3 working conference on data and applications security and privacy; 2010. p. 49–64.

[27] Baldi P, Baronio R, De Cristofaro E, Gasti P, Tsudik G. Countering GATTACA: efficient and secure testing of fully-sequenced human genomes. Proceedings of ACM CCS 2011;*11*:691–702.

[28] Naveed M, Agrawal S, Prabhakaran M, Wang X, Ayday E, Hubaux J-P, Gunter C. Controlled functional encryption. In: Proceedings of the 2014 ACM SIGSAC conference on computer and communications security; 2014.

[29] Jha S, Kruger L, Shmatikov V. Towards practical privacy for genomic computation. In: Proceedings of the 2008 IEEE symposium on security and privacy; 2008. p. 216–30.

[30] Chen Y, Peng B, Wang X, Tang H. Large-scale privacy-preserving mapping of human genomic sequences on hybrid clouds. In: NDSS'12: Proceeding of the 19th network and distributed system security symposium; 2012.

[31] Ayday E, Raisaro JL, Hengartner U, Molyneaux A, Hubaux J-P. Privacy-preserving processing of raw genomic data. DPM; 2013.

[32] Wang R, Wang X, Li Z, Tang H, Reiter MK, Dong Z. Privacy-preserving genomic computation through program specialization. Proceedings of ACM CCS 2009;09:338–47.

[33] De Cristofaro E, Faber S, Tsudik G. Secure genomic testing with size- and position-hiding private substring matching. In: Proceedings of the 12th ACM workshop on workshop on privacy in the electronic society; 2013.

[34] Ayday E, Raisaro JL, Rougemont J, Hubaux J-P. Protecting and evaluating genomic privacy in medical tests and personalized medicine. WPES; 2013.

[35] Kantarcioglu M, Jiang W, Liu Y, Malin B. A cryptographic approach to securely share and query genomic sequences. IEEE Transactions on Information Technology in Biomedicine 2008;12(5):606–17.

CHAPTER

Genealogical search using whole-genome genotype profiles

4

Yuan Wei[1], Ryan Lewis[2], Ardalan Naseri, PhD[2], Shaojie Zhang, PhD[1], Degui Zhi, PhD[2]

[1]*Department of Computer Science, University of Central Florida, Orlando, FL, United States;*
[2]*School of Biomedical Informatics, University of Texas Health Science Center at Houston, Houston, TX, United States*

1. Introduction

The year of 2018 will be noted, when looking back from the perspective of future humans, as the year that marked the beginning of traceable genetic genealogy. In April, the power of genetic genealogy has been dramatically demonstrated in front of the public by the arrest of the Golden State Killer, a horrific criminal who escaped law enforcement for 40 years. The critical evidence of this arrest was provided by autosomal genetic genealogical search, the process of comparing a person's DNA against a database with whole-genome DNA sequences of millions of individuals and identifying matches of a long streak of DNA markers that are indicative of traceable common ancestry.

The technical achievement genetic genealogy is due to the confluence of several technological developments and resource developments from the past two decades, including the completion of the human genome project, the accumulation of resources documenting human genetic variability, the affordability of genotyping for the mass, and the successful marketing of direct-to-consumer (DTC) genotyping services. Two developments that were central to genetic genealogical search are the availability of genetic databases of millions of individuals with their whole genome genotypes, and the efficient indexing and search algorithms for DNA matching. In this chapter, we are going to review these historic trends and basic concepts relevant to genetic genealogy using whole-genome genotype profiles. We will focus on the latest developments of the search algorithms that have impacted genetic privacy, and their societal implications.

2. History of personal genetic data

As the initial publication of the Human Genome Project in 2001 that mapped the overall base pair by base pair DNA sequence of the reference human genome, multiple government-funded efforts were developed to map variations in the human genome. Here, we review the development of these efforts.

Responsible Genomic Data Sharing. https://doi.org/10.1016/B978-0-12-816197-5.00004-8

2.1 HapMap

Beginning in October 2002, the International HapMap project set out to create a more complete haplotype map of the human genome. At the time of the project's infancy, the number of publicly documented single nucleotide polymorphisms (SNPs) was around 2.6 million. With collaborators from Canada, China, Japan, Nigeria, the United Kingdom, and the United States, the project led to the identification of 6 million more SNPs [1].

To successfully complete the task, samples from populations, who had been geographically separated, were collected and studied using different genotyping technologies. This allowed for a comparison of performance and cost between the technologies [2]. Sampling geographically separated populations was necessary to allow for HapMap to be useful to studies of all populations [3].

The data collected from this project were widely used in research until the development of the 1000 Genomes Project, and in June 2016, the HapMap resource was decommissioned [4].

2.2 1000 genomes Project

In light of the lesson learned from the HapMap project and taking advantage of improvements in genotyping technologies, the 1000 Genomes Project set out to find SNPs with allele frequencies as small as 1%. The project's main goal was "the development of a public resource of genetic variation to support the next generation of association studies relating genetic variation to disease" in September 2007 [5]. After all sequencing was finalized in 2015, the total number of samples collected for research was 2504. These samples came from 29 different populations [6]. The assessment of the data collected showed that any individual could find at least 95% of their accessible SNPs in the dataset [7]. The success of this project can be attributed to the utilization of different genotyping strategies and the wide diversity of the samples, allowing for the identification of a large number of variants (i.e. 88 million) [8].

2.3 UK Biobank and beyond

The UK Biobank is a well-known resource in the research community because it provides a rich data source to study all stages of the disease with the aid of personal genomic data. The program was established in 2006 and began collecting data on 500,000 volunteers from diverse backgrounds within the United Kingdom. Participants were between the ages of 40 and 69. What made this project unique from previous genomic research databases was the amount of personal and medical data attached to the genomic data. All types of samples were collected from the volunteers at the beginning of the project, and the participants agreed to have their ongoing healthcare tracked for future research. This profound initiative was funded by many health agencies in the United Kingdom, but the nonprofit charity allowed researchers from around the globe to access the data through a controlled access model [9]. As of June 2019, the UK Biobank has approved over 10,000 requests to access the database [10] with genetics/genotyping studies being the most approved area of interest at 1003 [11].

The success of the UK Biobank will continue to be recognized as time moves forward and as the participants continue to have their healthcare tracked. This novel approach to studying a wide range of topics in healthcare has inspired many other organizations to form and collect data in a similar manner as the UK Biobank. With the lessons learned from all of these research strategies, the future for personal genomics looks bright.

3. Direct-to-consumer genetic companies

One's genetic data is obviously valuable to oneself. Moreover, as one's genetic data does not change throughout one's life, it is more economical to conduct the genotyping test once and reuse its results for the rest of one's life. However, genetic testing was traditionally only reimbursable for conditions of clinical needs. When genotyping became affordable for a significant portion of the population, and as the public's awareness of the value of personal genetic data grew, the DTC companies began to flourish.

3.1 Early days

The DTC genetic testing industry has been around since 2000 when Family Tree DNA offered their first ancestry test. By 2008, there were around 24 companies in the United States offering similar services with approximately 460,000 people completing testing [11,12].

In the same year, Time magazine's invention of the year was the DTC DNA-testing service, 23andMe [13]. In 2007, 23andMe's service cost $999 and estimated a customer's genetic predisposition to health-related conditions by tracking well-studied SNPs. At the time of Time magazine's article, the price of their service had been reduced to $399. In the following years, the company went on to complete research studies on diseases that have an impact on many people, like breast cancer and Parkinson's, by using recently obtained customer samples, which numbered around 100,000 [14].

All the success of 23andMe did not go without notice from the United States government. On November 22 of 2013, the Food and Drug Administration (FDA) sent a warning letter to 23andMe to address the marketing strategy of the company. As the FDA believed 23andMe was providing a service whose purpose was to diagnose diseases or other conditions, it required the company to gain the FDA's approval before selling its product, which 23andMe had not previously done. In the letter, the FDA stated that the results 23andMe provided its customers concerning BRCA-linked genetic risk could result in serious health consequences if the test was a false positive or false negative [15].

3.2 Growth

With government regulations making it difficult to market and sell genetic tests designed to evaluate traits and conditions, the DTC industry was limited to providing genotype and ancestry data to its customers. One factor that aided in the growth of the DTC community during this time was the advancement of DNA sequence technology [16]. The development of whole-exome sequencing (WES) as a part of the next-generation technology allowed companies, like 23andMe, to lower the price of their genetic testing kits and grow the size of their databases. With help from financial investments, 23andMe was able to sell its kit at $99 starting December of 2012. The goal of the price reduction was to reach 1-million customers [14]. If not for the 2013 FDA warning letter, the milestone would have been reached earlier than the June 2015 date [17].

Furthermore, during this time of FDA restrictions, another DTC genetic testing company, AncestryDNA, emerged as a competitor to 23andMe. Starting in 2012, AncestryDNA was far behind in terms of customers served, but could provide an extra service that 23andMe could not when it came to ancestry data. Originally, AncestryDNA was Ancestry.com, who provided a service to their customers that allowed them to search historical documents and other evidence to build a genealogy tree. Combining this already well-established service with genetic testing for genealogical purposes was a winning combination, and the company reaches 1-million customers served by the end of 2015 [18]. The growth of AncestryDNA was unprecedented, seeing 350,000 new customers submitting genetic tests in the first 7 months of 2015 [17].

3.3 Current trends

23andMe's commitment to providing customers with genomic data detailing the customer's risk for certain health conditions forced them to work with the FDA and prove the validity of their test to sell their product for its intended purpose once again. The company took a small step in 2015 when it gained FDA approval for a genetic test to predict Bloom syndrome. As the condition is rare it did not attract a large market base [19]. In 2017, with the support of peer-reviewed research, 23andMe received FDA approval to sell a genetic risk test. The test provided information on 10 health conditions, most notably, Parkinson's disease and late-onset Alzheimer's disease [20]. Finally, in March 2018, the FDA approved the marketing of a product reporting BRCA mutations and the risk for different cancers, the same type of product that caused the initial warning letter in 2013. 23andMe was the first company to gain this approval for a cancer risk test [21].

As the FDA slowly gave approval to companies to market their products that provided information on health conditions, the number of customers signing up for the services increased. Aided with even further reductions to test kit prices, the top companies grew their customer base rapidly. As of early 2018, 23andMe stated that they had tested over 3 million customer samples, which was an explosion of growth since the 1 million marks in 2015. The one company that outpaced 23andMe in customer sales was AncestryDNA. Their marketing strategy of low price, $59, and access to the genealogy tools allowed them to reach an astounding 7 million customers by the beginning of 2018 [22]. Fig. 4.1 shows the rapid autosomal DNA database growth for each DTC company. The future of these companies is unknown, but the vast amount of genetic data collected holds great value to the future of genomic research.

3.4 GEDMatch and others

Many of the large DTC companies allow the customers to download their raw genetic data and use it at their own discretion. This practice has opened the door for third-party personal genomics databases and genealogy websites. With the FDA restrictions on what information could be marketed to customers, these services allowed the customer to explore their own genome without the aid of a DTC company. The most popular of these services is GEDmatch. Originally used to connect lost relatives with one another, in 2018 it made national headlines by being the main source of evidence in solving a long-running criminal investigation into the Golden State Killer [26]. After the exposure, the site had an increase in the number of genetic profiles uploaded, and by the end of 2018, they had close to 1 million profiles

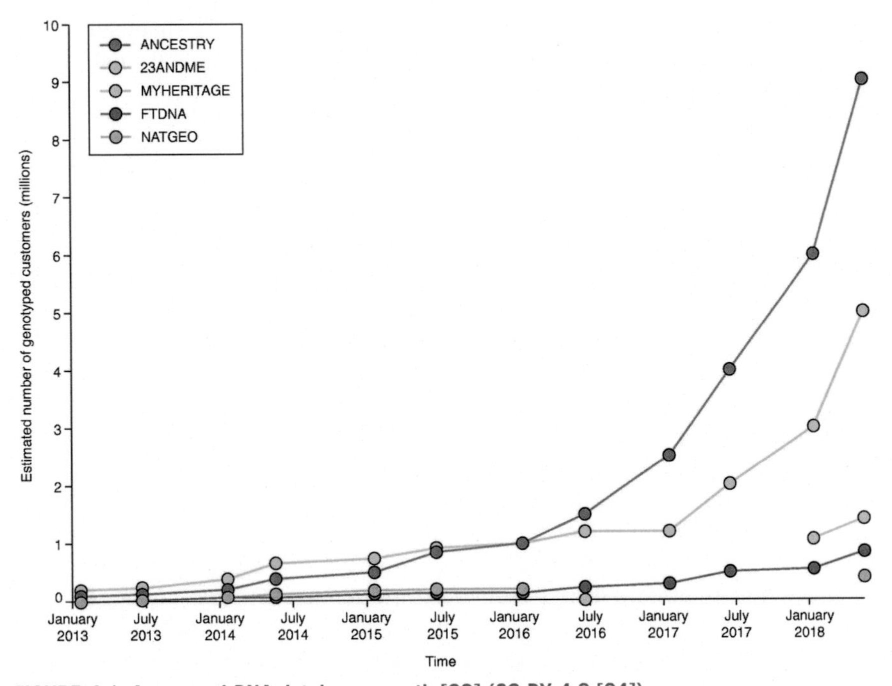

FIGURE 4.1 Autosomal DNA database growth [23] (CC BY 4.0 [24]).

In 2019, the number of people in the database has grown approximately 15 million for AncestryDNA, 10 million for 23andMe, 3 million for MyHeritage, and 1 million for FamilyTreeDNA [25].

in their database [27]. After this increase, the site began to sell memberships to the customers who wanted access to more tools used for genetic analysis.

A well-known nonprofit community website is DNA.Land. Different from GEDmatch, DNA.Land provides trait reports, and genetic profiles submitted to DNA. Land are not used in law enforcement investigations. The goal of DNA.Land is to collect genetic data for academic research, and in return for submission of genetic data, the site provides users with reports concerning their data. In October 2019, the site starts the transition into DNA.Land 2.0 and becomes a for-profit service similar to other third-party services [28].

As genetic testing increases in popularity, more third-party services like GEDmatch and DNA.Land will continue to enter the market to provide supplemental information to customers of DTC companies.

4. How to encode genotype information at the genome scale

Genotype information, as encoded by the sequence of four possible nucleotides, is inherently data in a very digital format. However, it is not efficient to encode all nucleotides because most of their positions are mostly unchanged in a population, and because the measurement of DNA variability is limited by current technology. In this section, we discuss the characteristics of genotype information and the data structures for encoding them.

4.1 Genotype

In biological terminology, an individual's genetic dataset is called a genotype. The word "genotype" was introduced by a geneticist, Wilhelm Johannsen, in 1909 [29]. He noted that an individual's genotype encompasses all of the individual's genes (fundamental units of hereditary information), which are inherited from both parents. This concept was first suggested by a geneticist, Gregor Mendel, in 1866 [30]. The term "gene" was introduced by Wilhelm Johannsen in 1905 [29]. It is made of deoxyribonucleic acid (DNA), (or ribonucleic acid (RNA)), which holds an individual's genetic information in a molecular form. The concept of DNA was identified by Francis Crick and James Watson in 1953 [31]. A DNA molecule contains two chains of nucleotides. Each nucleotide is comprised of sugars, phosphates, and one of the four nitrogenous bases: adenine (A), guanine (G), cytosine (C), and thymine (T). The sequence of bases along a strand of DNA determines an individual's biological information. The most common type of genetic variation in human DNA is single nucleotide polymorphisms (SNPs). (The other types of genetic variation are insertion, deletion of a DNA sequence, and structural variation.) An SNP is a difference in a single nucleotide of a DNA molecule between two individuals that occurs in more than 1% of the population [32]. SNPs are usually treated as tags within the chromosome, to identify the locations of genes on chromosomes.

For humans, genes are stored and encoded on each of the 23 pairs of chromosomes. The location of a specific gene on a chromosome is called a locus. The variant form of a gene at the same location for each pair of chromosomes is called an allele [33]. An individual's genotype for a specific gene is made of two alleles at the gene's locus on a pair of chromosomes (e.g., "AA" or "AT"). One allele is called the reference (major) allele (the most commonly observed nucleotide found at that locus in the entire human population) and another is called the alternative (minor) allele (the less commonly observed nucleotide found at that locus in the entire human population). It is rare but possible that there is a third type of nucleotide found in that locus. If both alleles of a gene are identical, we call such an organism *homozygous* with respect to that gene (e.g., both alleles are "A"). If the alleles are different, we call it *heterozygous* with respect to that gene (e.g., one allele is "A" and another allele is "T") [34]. The degree of similarity between two alleles of a gene determines the correlated biological trait of the organism that is inherited from the parent (e.g., eye color). The individual's phenotypic features are able to be identified by one's genotype, which consists of all alleles inherited from both parents.

The representations of human genotype information are pairs of alleles inherited from the two parents' chromosomes (one from each) (see Fig. 4.2). For instance, for a given SNP, the individual gets an allele "A" from its associated allele on the paternal side and it gets an allele "T" from its associated allele on the maternal side. The genotype representation of such a SNP is denoted as "AT" (or "TA", as most of the time we only know alleles are inherited from two parental chromosomes but we do not know wether each allele comes from the maternal or the paternal side). To process the genetic data efficiently, one conventional way is to represent the genotype of the SNP as a symbol. We assume that the reference (major) allele of the given SNP is "0" (e.g., "A") and the alternative (minor) allele of the given SNP is "1" (e.g., "T"). Then the genotype of the SNP can be represented by using one symbol: "0" if the reference allele "0" is found from both chromosomes at the locus of the SNP (e.g., two "A"s), "1" if the alternative allele "1" is found from both chromosomes at the locus of the SNP (e.g., two "T"s), or "2" if the reference allele "0" is found from one chromosome at the locus of the SNP and the alternative allele "1" is found from another chromosome at the locus of the SNP (e.g., one "A" and one "T"). An example of a genotype representation of three SNPs is: "AT|CC|CT", which is notated as "2|0|2" (assuming the reference allele of each SNP is "A", "C", and "C").

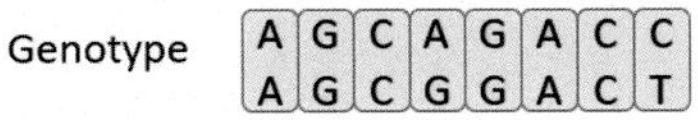

Genotype
A G C A G A C C
A G C G G A C T

FIGURE 4.2 An example of a genotype.

4.2 Haplotype

A haplotype is a collection of genes inherited from a single parent [35]. The term haplotype is derived from the word "haploid", which is having a single copy of each chromosome [36]. A haplotype containing only one set of chromosomes is the main difference between a haplotype and a genotype. Most genotypes are diploid cells, meaning 22 of 23 chromosomes are autosomes (having two copies of chromosomes, one from each parent). Only 1 of 23 chromosomes, the sex chromosomes X or Y, are haploid cells, which have one copy only from each parent. If an individual's sex chromosomes have one X chromosome and one Y chromosome, the X chromosome is copied from one's maternal side and the Y chromosome is copied from one's paternal side. If an individual's sex chromosomes have two X chromosomes, one of them is copied from one's maternal side and the other one is copied from one's paternal side.

The representation of human haplotype information is very similar to that of genotype discussed previously. Different from data in genotype form, one unit of data in haplotype form contains only alleles inherited from a single parent chromosome. In other words, it requires two units of data in haplotype form to store an individual's genetic information of a specific gene inherited from both parents (see Fig. 4.3), as it can be stored by one unit of data in genotype form. For instance, given an SNP, the individual gets an allele "A" from its associated allele on the paternal side and gets an allele "T" from its associated allele on the maternal side. The haplotype representation of such a SNP is "A" and "T" (comparing to the genotype representation: "AT" (or "TA")). Similar to the genotype representation, the allele on an SNP can be notated as a symbol: the reference (major) allele of the given SNP is "0" (e.g., "A") and the alternative (minor) allele of the given SNP is "1" (e.g., "T"). Although haplotype representation demands double storage as to the genotype representation, it provides additional information about the source of each allele. For example, we would know the allele "A" is from an individual's paternal side and the allele "T" is from an individual's maternal side if the given SNP data is in haplotype form. The unique parental haplotype information is very important as we need them to perform the search, because it usually is not obvious from the genotype representation.

4.3 Phasing

Phasing is a process used to convert an individual's genetic data from the genotype format into the haplotype format. Currently, the most economical method of human

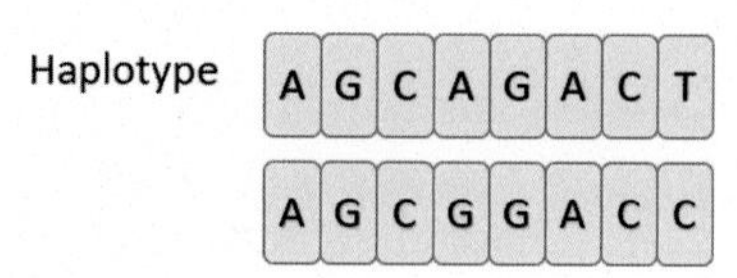

FIGURE 4.3 An example of haplotype.

DNA acquisition is using a DNA chip [37] to read extracted DNA samples by a set of probes to capture the nitrogenous bases detected on each probe. The output data is in genotype form and is stored in an SNP array. The SNP array is usually a sequence of alleles on variant sites from two copies of the individual's chromosomes. As an individual's genotype is a mix of two haplotypes from both parents, the nucleotide bases at each site are unordered, and it would not be possible to perform searches and find matches easily on such blended data (the search may result in a number of falsely matched segments). Thus we tend to infer two haplotypes from the given genotype (one is inherited from the paternal side and one is inherited from the maternal side, but it is possible that some alleles are mutated and not able to be found from either side). By translating a genotype into two haplotypes, we get to know the source of each SNP region (i.e., we get two inherited sequences of alleles: one represents the individual's paternal chromosome and the other represents the individual's maternal chromosome).

If an individual's family genetic data is accessible, the phasing process becomes trivial. We can compare alleles on each site from an individual's genotype to those from one's parental genotypes. If the allele matches one of the parental genotypes (or both parental genotypes if the parents have the same allele on the site), we know the source of the allele. We can deduce both of the individual's haplotypes in their entirety by comparing all the alleles. The phasing method using family information is called mother–father–child trios (both parents' genetic data is available) or parent–child duos (only one parent's genetic data is available).

Without the genetic data from an individual's family, we would have to speculate about the individual's haplotypes. This can be achieved with a statistical estimation on a large population because the DNA of all humans is very similar (about 99% of the whole genome) [38]. A reference population is a group of individuals who are not related to the given individual. The estimation is typically based on the computation of alleles' frequencies at each site from an unrelated population. Various phasing algorithms and programs based on this estimation have been proposed: PHASE [39], BEAGLE [40], IMPUTE2 [41], MaCH [42], SHAPEIT1 [43], SHAPEIT2 [44], Eagle [45], and Eagle2 [46]. They are mainly based on hidden Markov models (HMMs). In each iteration of a HMM, a model samples the haplotypes for each individual and treats them as the input for the next iteration. The probability of having a haplotype selected from the model depends on the number of times each allele is observed at each site. The allele frequency is calculated based on the assumption of Hardy–Weinberg equilibrium [47].

Many of the algorithms mentioned earlier improve over time through the phasing accuracy, the computational cost, and the capability of inferring a large number of datasets (e.g., they serve genome-wide association study (GWAS) [48]). Additionally, they also include imputation functionality (i.e., filling in missing values based on population) to construct the genetic data to be as close to the real data as possible (see Fig. 4.4).

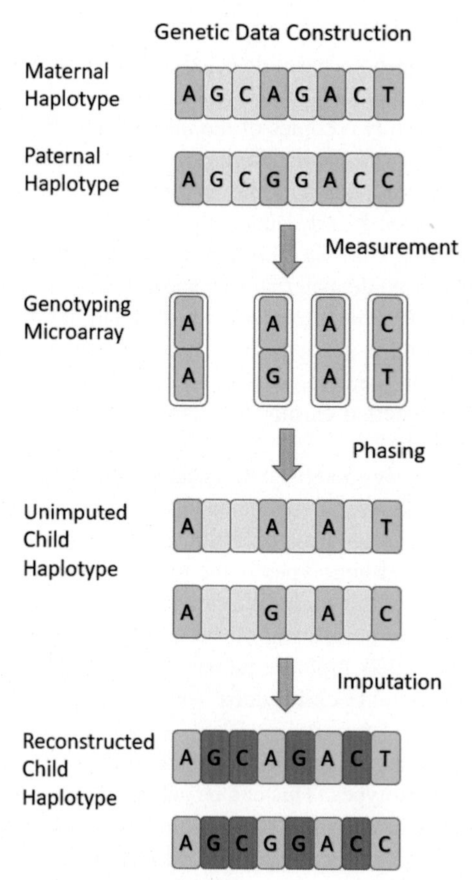

FIGURE 4.4 Genetic data construction diagram.

4.4 File format

Before DTC companies, genotypes for multiple individuals (e.g., from a research study cohort) were shared together and typically held information on one gene or a small genomic region, so the tabular format, for example, PLINK format, was popular. However, in the age of personal genomics, genotype files often contain genotype information across the entire genome. Genotype files can be very large, even for a single individual. DTC companies developed formats that facilitated the sharing of one individual's genotype information. In addition, now that sequencing is prevalent, there are many denser sets of genotype markers to be shared. So, the Variant call format (VCF) is being developed. The design of the VCF format makes strides toward a balance between human readability and machine readability (through gzip and tabix-index). However, the text-based file is not efficient for random access and subsetting. Furthermore, gzip does not offer efficient compression. Finding the most efficient genotype file formats is still an active research area.

4.4.1 VCF format

Variant call format (VCF) is a tab-delimited text file format storing gene sequence variations. It was originally developed to support the 1000 Genomes Project [49,50]. Currently, it is maintained by the Global Alliance for Genomics and Health (GA4GH) [51]. As the name suggests, it only stores the variations across the genome, rather than the entire genetic data. The design principle of VCF is to have a generic format for storing human genetic variations (SNPs, insertions, deletions, structure variants, etc.) alongside annotations (individual IDs, chromosome IDs, etc.). VCF is easily extensible to allow various genetic variations (i.e., the number of alternative alleles is larger than one). Additionally, although VCF is designed for computers to process the genetic data effectively, its representation is also friendly for humans to read.

Fig. 4.5 shows a sample VCF file containing human genetic data modified from the 1000 Genomes Project. The header part of VCF contains the definition of special keywords starting with a double pound sign "##" (e.g., ##fileformat = VCFv4.2), as well as the metadata starting with a single pound sign "#" (e.g., #CHROM). The metadata describes eight mandatory fields (guarantee unambiguity) and other optional fields (provide flexibility) for the body of the file. The eight mandatory fields are CHROM (the name of the chromosome), POS (the position of the variation on the given chromosome), ID (the identifier of the variation), REF (the reference allele at the position), ALT (the alternative allele(s) at the position), QUAL (the quality score associated with the alternative allele(s)), FILTER (a flag indicating if the position has passed the quality filters), and INFO (semicolon-separated key-

```
##fileformat=VCFv4.1
##FILTER=<ID=PASS,Description="All filters passed">
##fileDate=20150218
##reference=ftp://ftp.1000genomes.ebi.ac.uk//vol1/ftp/technical/reference/phase2_reference_assembly_sequence/hs37d5.fa.gz
##source=1000GenomesPhase3Pipeline
##contig=<ID=1,assembly=b37,length=249250621>
##...
##ALT=<ID=DEL,Description="Deletion">
##...
##FORMAT=<ID=GT,Number=1,Type=String,Description="Genotype">
##...
##INFO=<ID=AC,Number=A,Type=Integer,Description="Total number of alternate alleles in called genotypes">
##INFO=<ID=AF,Number=A,Type=Float,Description="Estimated allele frequency in the range (0,1)">
##INFO=<ID=AN,Number=1,Type=Integer,Description="Total number of alleles in called genotypes">
##INFO=<ID=NS,Number=1,Type=Integer,Description="Number of samples with data">
##INFO=<ID=DP,Number=1,Type=Integer,Description="Total read depth; only low coverage data were counted towards the DP, exome data were not used">
##INFO=<ID=VT,Number=.,Type=String,Description="indicates what type of variant the line represents">
##...
#CHROM  POS       ID           REF  ALT  QUAL  FILTER  INFO                                                      FORMAT  HG00096  HG00097  HG00099
22      16050075  rs587697622  A    G    100   PASS    AC=1;AF=0.000199681;AN=5008;NS=2504;DP=8012;VT=SNP        GT      0|0      0|0      0|0
22      16050115  rs587755077  G    A    100   PASS    AC=32;AF=0.00638978;AN=5008;NS=2504;DP=11468;VT=SNP       GT      0|0      0|0      0|0
22      16050213  rs587654921  C    T    100   PASS    AC=38;AF=0.00758786;AN=5008;NS=2504;DP=15092;VT=SNP       GT      0|0      0|0      0|0
22      16050319  rs587712275  C    T    100   PASS    AC=1;AF=0.000199681;AN=5008;NS=2504;DP=22609;VT=SNP       GT      0|0      0|0      0|0
22      16050527  rs587769434  C    A    100   PASS    AC=1;AF=0.000199681;AN=5008;NS=2504;DP=23591;VT=SNP       GT      0|0      0|0      0|0
22      16050568  rs587638893  C    A    100   PASS    AC=2;AF=0.000399361;AN=5008;NS=2504;DP=21258;VT=SNP       GT      0|0      0|0      0|0
22      16050607  rs587720402  G    A    100   PASS    AC=5;AF=0.000998403;AN=5008;NS=2504;DP=20274;VT=SNP       GT      0|0      0|0      0|0
22      16050627  rs587593704  G    T    100   PASS    AC=2;AF=0.000399361;AN=5008;NS=2504;DP=21022;VT=SNP       GT      0|0      0|0      0|0
```

FIGURE 4.5 A sample VCF file.

Data is from 1000 Genomes Project release phase 3 chromosome 22 The 1000 Genomes Project Consortium. A global reference for human genetic variation. Nature 2015:68–74. http://doi.org/10.1038/nature15393

value pairs describing additional information of the variation). One of the common optional fields is FORMAT (the data type of the given chromosome). For example, if "GT" is found in the FORMAT field, the data in the body of the file is in genotype format. Each allele is depicted as 0 (reference allele) or 1 (first alternative allele) or 2 (second alternative allele), etc., separated by either a forward slash "/" (unphased) or a vertical bar "|" (phased) [52].

VCF is used by various large-scale DNA sequencing projects and tools as their data type. The list includes (but is not limited to): IGSR (The International Genome Sample Resource) [53], ExAC (Exome Aggregation Consortium) [54], GATK (Genome Analysis Toolkit) [55], VEP (Variant Effect Predictor) [56], EVA (European Variation Archive) [57], dbSNP (Single Nucleotide Polymorphism Database) [58], UK10K (10,000 UK Genome Sequences Consortium) [59], and the NHLBI GO ESP (NHLBI GO Exome Sequencing Project) [50,60]. The most current version of full VCF specification can be found on the SAMtools website [52].

4.4.2 DTC format

The human genetic data formats used by DTC companies are very similar. They are all tab-delimited text files. They contain a header line indicating the column names, and the rest of the rows correspond to the data. Fig. 4.6 is an example. The design principle is to have customers be able to read and understand their genetic data easily while maintaining the ability for computers to process the data. Moreover, the tab delimited-based format is transferable from one DTC company to another, as customers may

#Genetic data is provided below as five TAB delimited columns. Each line
#corresponds to a SNP. Column one provides the SNP identifier (rsID where
#possible). Columns two and three contain the chromosome and basepair position
#of the SNP using human reference build 37.1 coordinates. Columns four and five
#contain the two alleles observed at this SNP (genotype). The genotype is reported
#on the forward (+) strand with respect to the human reference.

rsid	chromosome	position	allele1	allele2
rs587697622	22	16050075	A	A
rs587755077	22	16050115	G	G
rs587654921	22	16050213	C	C
rs587712275	22	16050319	C	C
rs587769434	22	16050527	C	C
rs587638893	22	16050568	C	C
rs587720402	22	16050607	G	G
rs587593704	22	16050627	G	G

FIGURE 4.6 A sample DTC file.

Data is from 1000 Genomes Project release phase 3 chromosome 22 individual HG00096 The 1000 Genomes Project Consortium. A global reference for human genetic variation. Nature 2015:68–74. http://doi.org/10.1038/nature15393.

like to export their raw genetic data from one company and import it to another company for further analysis, without retaking the DNA testing over and over.

The genetic data format that 23andMe uses is a tab-delimited text file. The file information is presented with the comments beginning with the pound "#" sign. Then, the file has a comment line containing four headers that describe the data: rsid, chromosome type, SNP position, and genotype. rsid (reference SNP cluster ID) is the SNP identification number. Its value is normally "rs" followed by a number. 23andMe uses the National Center for Biotechnology Information (NCBI) human reference genome (specifically, human genome assembly GRCh37 (build37)) [61] as the reference to SNP positions and DNA bases. The nucleotide bases used in their genotype data refer to the positive (+) DNA strand. Thus rsid is the indicator of the SNPs called as the human genome assembly GRCh37. For SNPs that are not reported in build37, they use internal id (its value is "i" followed by a number) as the identifier of their custom markers. The types of the chromosome (where SNP located) included in their genetic data are marked from "1" to "22" (nuclear DNA in autosomal chromosomes), "X" and "Y" (nuclear DNA in sex chromosomes), and "MT" (Mitochondrial DNA or mtDNA). The value of the SNP position is a number (in terms of Genome Reference Consortium build 37). The genotype is stored as a pair of unordered variants (e.g., CT). If a genotype for a particular SNP cannot be confidently identified, the value used is two dashes, "–" as the entry of the genotype, instead of a two-letter genotype. For the body of the file, each line corresponds to a single SNP, providing information regarding the four headers [62].

The genetic data format of AncestryDNA is a tab-delimited text file. Similar to 23andMe, the file starts with the comments and header descriptions (beginning with the pound "#" sign). The rest of the file shows the SNP data for each subsequent line. Each SNP has five fields of data: rsid, chromosome, position, allele1, and allele2. The data file uses human reference build 37 (the current major release) as the reference of the SNP information. Different from 23andMe, AncestryDNA does not provide any custom markers in the file. rsid (reference SNP cluster ID) is the identification number of the examined SNP. The types of chromosomes (where SNPs are located) are all numbers (i.e., "1" to "22" correspond to chromosomes 1 to 22; "23" corresponds to chromosome X, "24" corresponds to chromosome Y, "25" corresponds to chromosome XY, and "26" indicates MT, which corresponds to SNPs detected on mtDNA). The value of the SNP position is a number (in terms of Genome Reference Consortium build 37). The fields allele1 and allele2 store two alleles observed at the SNP location on the forward (positive) strand of the chromosome as its unordered genotype. The value can be the same or different (i.e., "C" and "C", or "T" and "C"). If the alleles cannot be determined, the values of allele1 and allele2 are both set to "0" [63].

The genetic data format of MyHeritage is a tab-delimited text file. The file contains a header line beginning with the pound "#" sign describing the genetic data, and lines that each have five fields of SNP data: rsID, Chromosome, Position, Allele1, and Allele2. The rsID field holds the reference SNP cluster ID, indicating the name of the variant SNP. The Chromosome field holds the chromosome identification number for the 23 pairs of chromosomes (i.e., "1" to "22" and "X" and "Y"). MyHeritage does not provide SNP data observed from mtDNA. The Position field holds the number indicating the SNP location on the chromosome (in terms of Genome Reference Consortium build 37). The fields of Allele1 and Allele2 hold a pair of letters indicating the observed genetic variants on the forward strand of the chromosome with unidentified order (e.g., "C" and "C", or "T" and "C"). If the variant type is indel (insertion or deletion of nucleotides), the letter "I" (for insertion) or "D" (for deletion) is used in this field. A dash "-" is used if the variant is not determined [64].

The genetic data format, "Family Finder", from FamilyTreeDNA (FTDNA) is a comma-separated variable (CSV) file and all fields are double quoted. The first row of the file indicates the name of the four data columns: RSID, CHROMOSOME, POSITION, and RESULT. The actual genetic data start from the second row of the file. Each data row contains values corresponding to the four data columns. The RSID field stores the reference SNP cluster ID corresponding to the National Institutes of Health (NIH) dbSNP database [58]. The field of CHROMOSOME stores the examined chromosome name ("1" to "22" for autosomal chromosomes, and "X", "Y" for sex chromosomes). The field of POSITION stores the location identification number of the specified chromosome of the SNP. The SNP coordinates (both chromosome name and SNP position) follow genome assemblies build 37. The field of RESULT stores two-letter allele values observed from the SNP (e.g., "CC" or "CT"). If the variant type is indel (insertion or deletion of nucleotides), the letter "I" (for insertion) or "D" (for deletion) is used in this field (i.e., "II," "DD," "DI"). If the result is not clear, two dashes "–" are used in this field. The RESULT field always shows doubled values (even for males who only have one X chromosome). The FTDNA "Family Finder" test does not include SNPs observed in mtDNA or the Y chromosome, though they are available via different types of tests [65].

4.4.3 Compressed format

4.4.3.1 PLINK format (BED)

PLINK is a genetic data toolset developed by Shaun Purcell et al. [66]. The goal of PLINK format is to handle large datasets and perform analysis on them in a computationally efficient manner. The way PLINK manages data is to have data represented in a compressed fashion, with a binary text-based transformation tool attached. To provide whole-genome association analysis, PLINK offers complimentary tools for summary statistics (e.g., calculating allele frequencies), population stratification (e.g., identifying outlying individuals), association analysis (e.g., disequilibrium test), and identical by descent (IBD) estimation (e.g., estimating the genome-wide level of relatedness between individuals). PLINK Version 2.0 was updated by Christopher Chang et al. [67], with improvements to performance and compatibility.

PLINK supports two types of genetic data formats: flat files and binary files. The flat-file format is the standard PLINK format and it consists of a PED file (storing individual phenotype and genotype data) and a MAP file (storing marker description data) [68]. Both of them are whitespace (space or tab) delimited files. For each line in a PED file, it contains six mandatory fields: family id, individual id, paternal id, maternal id, sex (1 = male; 2 = female; other = unknown), and phenotype (a quantitative trait or an affection status; 0 = unknown; 1 = unaffected; 2 = affected). The genotype data of the individual start from the seventh field of each line in a PED file (2 space or tab-separated fields consist of one SNP; 0 = missing). For each line in a MAP file, it contains four mandatory fields: chromosome name (1–22, X, Y, or 0 if unplaced), rsid (reference SNP cluster id), genetic distance in centiMorgan (cM), and base pair physical position (bp units). The total number of rows (i.e., the total number of SNPs reported) in the MAP file should match the doubled total number of the fields starting from the seventh field in the correlated PED file because, in the PED file, each line represents an individual and each individual should have two allele values per SNP. The binary file format is an efficient PLINK format, and it consists of a BED file (storing individual genotype data), a BIM file (storing marker description data), and a FAM file (storing individual phenotype data) [69]. Both BIM file and FAM file are whitespace (space or tab) delimited files. For each line in a BIM file, it contains six mandatory fields (as an extended MAP file): chromosome name (1–22, X, Y, or 0 if unplaced), rsid (reference SNP cluster id), genetic distance in centiMorgan (cM), base-pair physical position (bp units), first possible allele of the specified SNP, and second possible allele of the specified SNP (value can be the same or different from the first one). For each line in FAM file, it contains six mandatory fields: family id, individual id, paternal id, maternal id, sex (1 = male; 2 = female; other = unknown), and phenotype (a quantitative trait or an affection status; 0 = unknown; 1 = unaffected; 2 = affected).

A BED file is a compressed file. It starts with a 2-byte (16-bit) magic number "01101100 00011011", indicating the file format is BED. Then the third byte (8 bits) indicates how data is presented in the specified BED file. "00000001" indicates that the BED file lists all individuals for each SNP, whereas "00000000" indicates that the BED file lists all SNPs for each individual. Starting with the fourth byte, each byte (8 bits) can represent up to four genotypes (i.e., 2 bits represents a genotype). The genotypes are encoded in reverse order in a byte (8 bits) (the eighth and the seventh bits indicate the first genotype in the specified byte, the sixth and the fifth bits indicate the second genotype in the specified byte, etc.). The rules for how two bits represent the genotype are as follows: "00" indicates that the genotype is *homozygous* and the identical allele name is the first allele of the specified SNP in the BIM file; "11" indicates that the genotype is *homozygous* and the identical allele name is the second allele of the specified SNP in the BIM file; "01" indicates that the genotype is *heterozygous* and the two different alleles' names are the first and the second allele of the specified SNP in the BIM file (unordered); "10" indicates the genotype information is missing. PLINK is aware of the total number of SNPs and individuals because it can get the information from the corresponding BIM file and FAM file. When PLINK reaches the end of the SNPs (or the end of the individuals) while it is reading the BED file, it skips the remaining bits in the byte and starts reading from the next byte.

4.4.3.2 UK Biobank format (BGEN)

BGEN is a data format storing either typed or imputed genotype data with the imputed genotype probability (i.e., dosage value). It was developed by Gavin Band and Jonathan Marchini [70]. The UK Biobank [71] uses BGEN as the format of genome-wide imputed genotypes [71,72]. Other projects, including Wellcome Trust Case Control Consortium 2 [73], the Malaria Genomic Epidemiology Network (GEN) [74], and the Avon Longitudinal Study of Parents and Children (ALSPAC) [75], also use BGEN format [76].

BGEN is the file format of a compressed binary GEN file. GEN is an Oxford text genotype file format which is the output of IMPUTE, one of the popular genotype imputation methods [41] and it stores unsigned integers in little-endian order (least significant byte first) [77]. A BGEN file starts with an offset block (4 bytes (32 bits)), indicating the location of the first variant data block (i.e., the offset value equals the number of bytes used for the offset block itself plus the number of bytes used for the header block). The header block (following the offset block) contains the number of samples, the number of variant data blocks, and the flags. The flags define if SNP blocks are compressed or not, if sample identifiers are stored or not, the layout of SNP blocks (biallelic or multiallelic), and unspecified fields reserved for future use. If the sample identifier flag is on, a sample identifier block is the next one after the header block; otherwise, the variant data blocks are the next ones. A sample identifier block stores a list of single identifiers for each sample. A variant data block stores the SNP data identifier and the genotype probability. The SNP data identifier includes the variant identifier (alternate id) (e.g., chip manufacturer ID for assayed SNP), rsid (reference SNP cluster id), the chromosome name, the SNP position, the number of alleles, first allele's name, second allele's name, third allele's name (if any), etc. [78]. The first allele is the reference allele on the forward DNA strand [79]. The genotype probability data section stores the probabilities of the permutation of all possible genotypes that the sample individual may have at the SNP (e.g., there are three genotype possibilities for a biallelic SNP: having two "A" alleles, having one "A" allele and one "G" allele, and having two "G" alleles). The genotype probability is encoded in bit representation and each of them takes 2 bytes of storage space.

4.4.3.3 GDS format

Xiuwen Zheng et al. propose an array-oriented data format to store genome-wide data, named Genomic Data Structure (GDS) [80]. GDS data format is designed for multicore computing mechanisms. Each byte encodes four SNPs (i.e., 2 bits per SNP). This makes GDS efficient when running on any multicore computer architecture as any four SNPs can be processed simultaneously. On top of the GDS format, Xiuwen Zheng et al. introduce a sequence array (SeqArray) data format [81]. The SeqArray data format has the header section and the data field section for each variant. The data field section has a few subsections, which contain information on sample id, variant id, position, chromosome, reference and alternate alleles, genotype data block, phasing states, and annotation. The genotype data block contains a number of subblocks with 2-bit arrays in compressed format. Each compressed subblock is stored sequentially and the last subblock is the index table. The index table contains two arrays, storing the size of compressed data of each subblock and the size of raw data of each subblock. It is used by the storage algorithm to compress and decompress the data in each subblock. GDS and SeqArray file formats are implemented in the R programming language and can be found in the Bioconductor package (Bioconductor is a bioinformatics open-source software project) [82,83]. The package also provides tools to convert other popular genotype data, such as VCF, into GDS format. GDS format is primarily adopted by National Heart, Lung and Blood Institute Trans-Omics for Precision Medicine (NHLBI TOPMed) [84,85].

4.4.3.4 Population-genetics inspired format: PBWT (BGT) and tree sequence (tsinfer)

More advanced genotype formats leverage the population genetics processes that shape the genetic variabilities of the population. Positional Burrows–Wheeler Transform (PBWT), introduced by Richard Durbin, is a data structure based on positional prefix array [86]. PBWT is designed for efficiently processing haplotype data using linear time and space. It takes advantage of the haplotype feature that adjacent rows have strong correlations, which is the result of linkage disequilibrium (LD). Linkage disequilibrium postulates the correlation between nearby alleles is stronger than that between distant alleles within a population (i.e., the nearer alleles have a higher chance to be inherited together (or having lower probability of crossover between them) during the reproduction for a specific species; thus the individuals among the population are likely to have similar regions) [87]. BGT, a PBWT based data structure, is a compact data format proposed by Heng Li, that is aimed at for efficiently storing and querying whole-genome genotypes [88]. It stores the genotype data and the sample annotation data in separate files (as VCF stores all data in one big file). The genotype data is stored as a 2-bit integer matrix, where the row represents the pair of alleles and the column represents the sample. BGT arbitrarily phases the genotypes into haplotypes. Thus each sample has two columns, representing two haplotypes. For each column, if the allele matches the reference allele of the row, its value is set to "0"; If the allele matches the alternative one of the row, its value is set to "1"; if the allele is neither the reference one nor the alternative one of the row, its value is set to "3"; if the allele is unknown, its value is set to "2." The matrix is stored as two PBWT data structures (i.e., all the lower bits are stored in one PBWT and all the higher bits are stored in the other PBWT). The reason that the BGT format adopts PBWT is to utilize the linkage disequilibrium feature. With the LD feature, BGT is able to improve the compression ratio by using the run-length encoding (RLE, which stores the data value as a single value and the counts of the same data value occurring consecutively as the other value, to save the storage space) [89]. The sample annotation data is stored in a flat metadata format (FMF). It is a tab-delimited text file, and it contains sample phenotype data and SNP annotation data. Each row stores a meta-annotation starting with the individual's name, and then values with a custom-defined "key:type:value" format. The type section takes "z" as a string, "f" as a real number, and "i" as an integer. The benefit of having an FMF file separated from the genotype data file is that the FMF file can be reused across multiple BGT files, and it avoids reprocessing the genotype data each time when the sample annotation is being updated. Additionally, having two separate files can allow for querying genotype data in a private manner (i.e., no need to access the FMF file, which makes sample individuals unidentifiable) [90].

The tsinfer (tree sequence inference) is a data structure proposed by Kelleher et al. [91], with tskit (tree sequence toolkit). The goal of introducing tsinfer is to represent human genetic data in a tree structure, with ancestry information embedded. The structure assumes the ancestral allele for all sites is zero "0" and the derived allele (when a mutation happens) is one "1." Each node represents an individual, and each edge stores the mutation information of all sites of the individual. Each branch in the tree indicates the occurrence of a recombination (crossover) event. Compared to the conventional matrix representation (e.g., VCF format), the tree sequence format is more scalable. It reduces the space complexity from $O(mn)$ to $O(m + n)$, where m is the number of sites and n is the number of individuals. This is due to the tree sequence only storing the mutation information (i.e., when the value of each site changes from the ancestral allele to the derived allele). Taking advantage of the tree sequence data structure, tsinfer can easily obtain genetic history information and make statistical inferences from a large population dataset.

The currently existing genetic data file formats are developed for different purposes (see Table 4.1). As the study of population genetics rises, more efficient, compressible, and goal-specific data formats may emerge in the future.

5. Identity-by-descent segment and familial relatedness

The continuous chunk of autosomal DNA (chromosome) of an ancestor will be broken into pieces by the natural force of meiotic recombinations. As a result, closely genetically related individuals may share continuous segments of DNA that are identical by descent (IBD). IBD segments hold critical information for inferring familial relatedness in genealogical search.

5.1 Genetic distance

Genetic distance is a genetic divergence measurement between either species or populations within a species [92]. For autosomal DNA comparisons, genetic distance refers to the length of the shared DNA segment in centiMorgans (cM) [93]. A centiMorgan (also genetic map unit (mu) [94]) is a unit of measure used to approximate genetic distance along chromosomes. The name was coined by the geneticist Thomas Morgan and his student Alfred Sturtevant [95]. A genetic distance is not a physical distance but an implied probability of a crossover occurring along the distance between loci on a chromosome, while a megabase (Mb) is the unit used to

Table 4.1 A summary of genetic data file formats.

Genetic data format	Compressibility	Extensibility	Human readability	Field of use	Design feature
VCF	No	Yes	Yes	Academia	The generic format for genetic variations and annotations
DTC format	No	No (individual data per file)	Yes	Industry	Compatibility for all DTC companies
PLINK	Yes	Yes	No	Academia	Analysis of large datasets
BGEN	Yes	Yes	No	Academia	Analysis of imputed genotype probabilities
GDS	Yes	Yes	No	Academia	Efficiency on the multicore computing systems
BGT	Yes	Yes	No	Academia	Efficiency on data storage
Tsinfer	Yes	Yes	No	Academia	Analysis of the genetic history

measure the physical distance. In a human organism, one single centiMorgan corresponds to approximately 1 million base pairs (bp) (or 1 megabase) [96]. The centiMorgan unit is used to quantitate crossover frequencies, and 1 centiMorgan is considered equivalent to a crossover frequency of 1% of a marker that is separated from another marker on a DNA segment in a single generation [97]. Currently, all biotechnology companies use centiMorgan to denote the estimated size of matching DNA segments in their autosomal DNA tests [98].

5.2 What is IBD

Identical by descent (or identity by descent) (IBD) is a biological terminology proposed by Gustave Malécot [99,100]. Originally, it serves as an indication of two homologous alleles descending from a common ancestor. The probability of two alleles being IBD is calculated with a reference population, which is from a known pedigree. On top of the traditional analysis of IBD, John Kingman established coalescent theory, which is a theoretically based genealogy model. In coalescent theory, all alleles are considered being descended from one common ancestor randomly at different times [101]. Taking advantage of the HapMap project [102], it is possible to use only genetic markers (SNPs) to estimate the probability of two alleles being IBD without a reference population [103]. Contemporarily, from the genome-scale perspective, IBD is redefined as two homologous chromosome segments being inherited from a common ancestor [104]. A matching IBD segment is described as all the alleles on a DNA segment of a chromosome inherited from either paternal or maternal side being identical [105]. Fig. 4.7 shows the IBD segments that a child

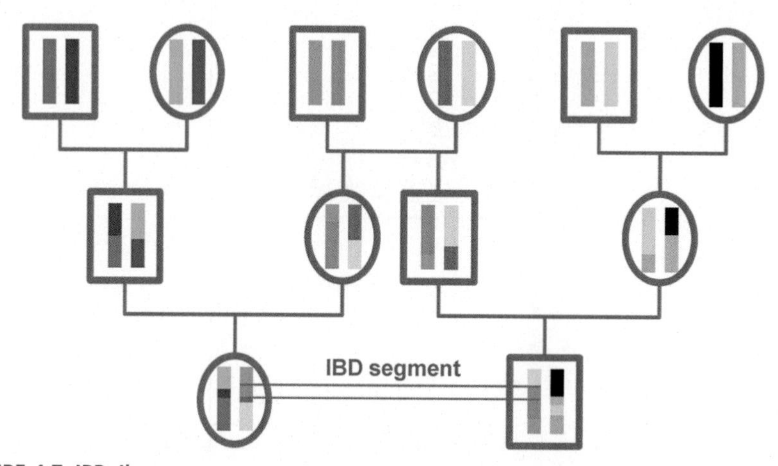

FIGURE 4.7 IBD diagram.

The female individual (circle shape) in the bottom row shares an IBD segment (orange color) with her first cousin, the male individual (box shape) in the same row. This shared IBD segment (orange color) is inherited from their grandfather (the male individual (box shape) in the middle of the top row) [109] (CC BY-SA 3.0 [110]).

inherits from one's parents. Note that IBD is different from identical by state (or identity by state) (IBS). IBS describes two DNA segments that are identical, but they have no genealogical relevance [106]. In general, the shared segments (having identical alleles) are IBS, and IBS can be either IBD (if the shared segments are copied from a common ancestor), or a false-positive segment. The false-positive segment (also pseudo segment) can result from mutations, phasing errors, lack of phasing, etc. [107]. The false segment caused by the lack of phasing is called identical by chance (or identity by chance) (IBC). The IBC segment is caused by having a false match on unphased genotypes (i.e., an individual's matching segment zigzags back and forth between one's parents' segments) [108].

To determine if IBS segments are IBD segments, the ancestral lines of the matching segments need to be traced back, to find the time when the matching segments coalesce. Doug Speed and David Balding [111] perform a simulation of IBD segment relatedness, where they demonstrate the distribution of the generation of the common ancestor in the past depending on the lengths of IBD segments. There is about a 50% probability that IBD segments in the range of 5 −10 cMs may come from a common ancestor 20 generations back in time. The longer the lengths of IBD segments, the higher the probabilities are that they are from a common ancestor, from a few generations back in time.

For genealogical search, each biotechnology company sets up its own minimum IBD segment thresholds as DNA matching criteria (they use it combined with total lengths of IBD segments and other criteria to infer the familial relatedness). The confidence autosomal thresholds of each company are as follows: 7 cMs (or 700 SNPs) for 23andMe [112]; 8 cMs for MyHeritage [113]; 6 cMs for AncestryDNA [114]; and 7.69 cMs for FamilyTreeDNA [115]. The value of the minimum IBD segment threshold, however, may not be set small enough for the matching IBD segment search. If the value of the IBD segment is tiny, there is a high chance that the IBD segment is a false-positive segment (the false-positive rate of 2–4 cM segments is over 67% [116]). The small IBD segment could come from a common ancestor, a large number of generations back in time, and genealogical information of such an ancestor may not be available to be identified. Another case is that the small IBD segment could be identical by population (or identity by population) (IBP) segment. IBP segment is an identical DNA segment found in multiple ancestors. It is commonly seen in the endogamous population. As it is a popular segment shared by a population, it is extremely difficult to trace the ancestral line back to find the common ancestor. IBP is usually considered as a noise segment [108].

5.3 Cousin nomenclature

In general, family members are classified mainly based on gender (sister vs. brother, mother vs. father, aunt vs. uncle, niece vs. nephew, etc.), generation (parent, grand parent, great grand parent, second great grand parent, third great grand parent, etc.), and consanguinity (person vs. parents (or children), vs. grand parents (or siblings, or grand children), vs. great grand parents (or aunts/uncles, or nieces/nephews, or great grand children), vs. second great grand parents (or great aunts/uncles, or first cousins, or grand nieces/nephews, or second great grand children), etc.). According to Black's Law Dictionary (which is the standard reference for defining family relationships [117]), cousin is defined as a child of the person's aunt or uncle [118]. Specifically, cousin is the one who shares a common ancestor with the person, and such a common ancestor is at least two generations away. The degree of the cousin relationship can be formally defined by using the ordinals of the cousin-ness (first cousin, second cousin, third cousin, etc.) [119] and the ordinals of the generation (the first cousin once removed, the first cousin twice removed, the first cousin three times removed, etc.). The ordinal of the cousin-ness indicates the distance of the closest common ancestor (i.e., the number of generations away from the closest shared common ancestor) between the person and one's cousin, who is from the same generation.

For example, if the closest common ancestor that the person and someone who is from the same generation of the person share is two generations away (i.e., the person's grand parent), someone is the person's first cousin. If the person and someone who is from the same generation of the person share a closest common ancestor who is three generations away (i.e., the person's great grand parent), someone is the person's second cousin. The ordinal of the generation indicates the distance of the generations between the person and one's cousin, who is from a different generation. For example, if someone is the person's first cousin but who is separated from the person by one generation, someone is the person's first cousin once removed (also one generation removed). The generation difference can go either direction in the person's family relations chart. In the previous example, both the person's first cousin's child and the person's second cousin's parent are the person's first cousins once removed. In other words, first cousins are the person's nonsiblings who share grand parents, and first cousins once removed are the person's relatives who share first cousin relationship being one generation removed [120]. Fig. 4.8 is the consanguinity chart visually describing the relationship among relatives.

5.4 How IBD is related to family relationships

Based on Mendel's principles of inheritance (particularly, the law of segregation of genes), an offspring inherits a pair of alleles from one's parents (one allele from each parent) [122]. From biology perspective, the underlying process is called meiosis, a cell division that reduces the chromosome number by half, to produce four haploid gametes (i.e., a cell provided by each parent and later joins with a cell of the opposite sex to form a zygote (i.e., a single cell later develops into an individual person)), which contain one set of 23 chromosomes. Later on, two gametes fuse into one diploid zygote, which contains 23 chromosomes from both the paternal and maternal side [123]. The recombination (also crossover) in the process of meiosis occurs at a rate of 0.01 cM, which causes the nearby chromosomal locations to be strongly positively correlated, that is, forming a possible IBD segment [104]. As biologically an

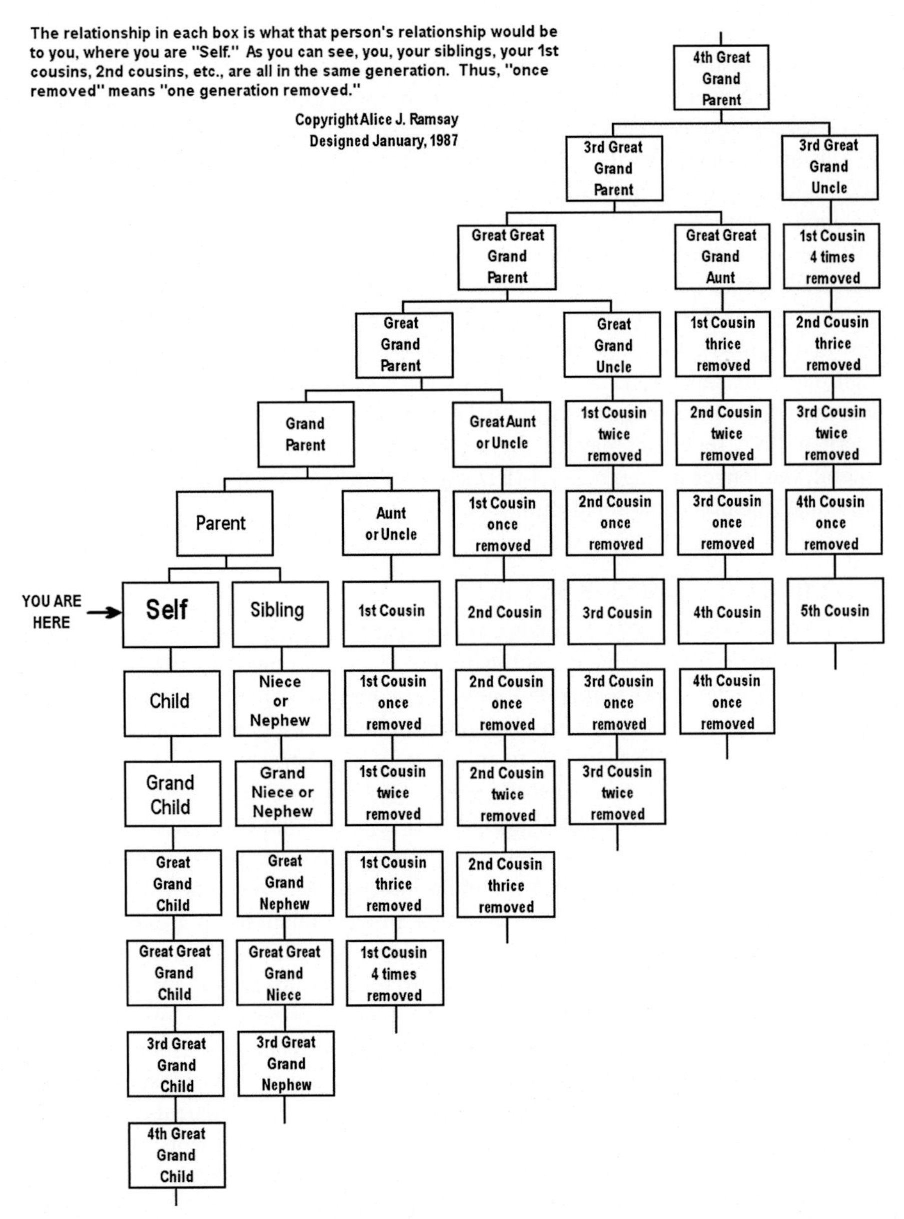

FIGURE 4.8 The consanguinity chart (CC BY-ND 2.5 [121]).

individual inherits half of the chromosomes from each parent (one's DNA segments are inherited equally from one's parents (i.e., 50% DNA segments from each parent)), and the allele locations on chromosomes are strongly positively correlated (the inherited DNA segments from either single parent are nearby), it is plausible that the individual shares approximately 50% of one's IBD segments with one's parent (with respect to either mother or father). Libiger and Schork obtain 49.8% as the result of above, from a simulation of computing the probabilities associated with shared chromosomal segments [124]. It is not difficult to make a further inference using the degree of the relationship to obtain the approximate percentages of IBD segments that the individual shares with other family members. For example, the approximate percentages of IBD segments the individual shares are 100% with one's identical twin, 50% with one's sibling, 50% with one's child, 25% with one's grand parent, 25% with one's aunt or uncle, 25% with one's niece or nephew, 25% with one's grand child, 12.5% with one's great grand parent, 12.5% with one's grand aunt or uncle, 12.5% with one's first cousin, 12.5% with one's grand niece or nephew, 12.5% with one's great grand child, etc. [125]. Note that even though the percentage of shared IBD segments may look the same, the actually inherited DNA segments may be slightly different for each pair of individuals. In general, the longer the aggregated IBD segment an individual shares with another individual, the closer the relationship they have.

5.5 Expected IBD family sharing

With two individuals' DNA segments, it is possible to predict their relationship by comparing their DNA segments and using the result information, such as the total number of matched IBD segments found, the aggregated amount of matched IBD segments found, or the longest length of the matched IBD segments found. The total number of matched IBD segments to each family member, however, is not conclusive. Previously, it was shown that the percentage of shared IBD segments implies the family relationship. The actual shared IBD segments, however, may have intersections among multiple family relationship groups. In other words, each relationship group has a range of its corresponding matched IBD segments (measured as genetic distances in centiMorgans). The range of each relationship group has overlapped with that of other groups, which makes the relationship prediction very tricky. According to AncestryDNA's simulation result, relationship groups are categorized as the number of meiosis occurrences: one meiosis occurrence corresponds to parent, child, or sibling relationships, two meiosis occurrences correspond to grand parent, aunt (or uncle), niece (or nephew), or grand child relationships, etc. [114]. Each relationship group has its correlated range of total IBD segments, indicating that the individual who has such a number of matched IBD segments may belong to the group. The range is calculated based on the probability distribution of the total number of IBD segments for each relationship group (or the number of meiosis occurrences). Table 4.2 shows the percentage, average, and range of shared IBD segments correlated to the detected relationship in general (various biotechnology companies may use a slightly different genetic distance of shared IBD segments as the threshold for detecting relationships).

Besides the overlapped range of shared IBD segments, the probability of finding shared IBD segments decreases as we expand the relationship further in an

Table 4.2 The average and the range of shared IBD segments per relationship.

Relationship group (cluster)	Average shared IBD (percentage)	Range of shared IBD (percentage) [112]	Average shared IBD (centiMorgan)	Range of shared IBD (centiMorgan) [126]
Identical twin	100%	Not available	6800.00 [127]	Not available
Parent, child	50%	Variable	3400.00	Variable
Sibling	50%	Variable	2550.00 [125]	2209.00–3384.00
Grand parent, aunt (or uncle), niece (or nephew), grand child	25%	Variable	1700.00	1294.00–2230.00
Great grand parent, great aunt (or uncle), first cousin, great niece (or nephew), great grand child	12.5%	7.31%–13.80%	850.00	486.00–1761.00
second great grand parent, great grand aunt (or uncle), frst cousin once removed, great grand niece (or nephew), second great grand child	6.25%	3.30%–8.51%	425.00	131.00–851.00
Third great grand parent, second great grand aunt (or uncle), first cousin twice removed, second cousin, second great grand niece (or nephew), third great grand child	3.125%	2.85%–5.04%	212.50	47.00–517.00

individual's family relationship chart. Kevin Donnelly proposed a mathematical model showing the theoretical probabilities of detectable DNA segments in genealogical relationships. The empirical formula to calculate the probability of having no detectable shared DNA segments is as follows: $e^{-\frac{kL}{2^k}}$, where L is the chromosome length in kilobase (kb, equal to 1000 base pairs). The total autosomal chromosome length in kb is about 33. k is the coefficient indicating the cousin relationship to the individual: $k = 2s + t$, for sth cousins t times removed [128]. From the equation, we derive the relationship table of the probability of having detectable shared DNA segments from cousins, alongside those from biotechnology companies (we only consider cousin relationships because the genetic data of other family members who are several generations away may not be applicable in practice).

Table 4.3 shows the probability of having a detectable shared DNA segments in both theory and practice. As we see from Table 4.3, the possibility of an individual's cousins having shared IBD segments decreases as the relationship distance grows, roughly following an inverted sigmoid curve. It is often challenging to find distant cousins as their DNA segments are not likely traceable from either a theoretical or empirical perspective.

6. Genealogical search

6.1 What is a genealogy search

Genealogy search involves constructing a family tree, locating relatives, and sometimes providing historical records to the customer. The genealogy report usually contains an individual's family-oriented information, as the ancestry report focuses information on an individual's origins (i.e., the relationship between an individual and the entire human population).

Table 4.3 The probability of having a detectable shared DNA segments in theory and in practice.

	Probability of having detectable shared DNA segment			
		In practice [129]		
Cousin relationship	In Theory [128]	By 23andMe	By AncestryDNA	By FamilyTreeDNA
First cousin	100.00%	100.00%	100.00%	100.00%
Second cousin	100.00%	100.00%	100.00%	99.00%
Third cousin	97.70%	89.70%	98.00%	90.00%
Fourth cousin	69.30%	45.90%	71.00%	50.00%
Fifth cousin	30.20%	14.90%	32.00%	10.00%
Sixth cousin	10.10%	4.10%	11.00%	2.00%

6.1.1 Difference between ancestry and genealogy

An ancestry report can tell an individual their estimated ethnic/geographic origins going back hundreds to thousands of years [130]. To be able to derive the estimates for the individual, a reference panel must be established. The reference panel is composed of samples whose ethnic/geographic origins have been confirmed. Confirmation can be done through genealogy analysis or samples from previously annotated data. One consideration when developing a reference panel is ensuring the number of related family members is minimal because too many closely related individuals in the panel could cause a bias in the results. In addition, sample outliers or samples that do not fit in with their confirmed ethnic/geographic group must be removed. To do this, the results of a principal component analysis (PCA) can be used to identify these samples. A PCA takes hundreds of thousands of related dimensions, or single nucleotide polymorphisms (SNPs), from the samples, and reduces them down to uncorrelated components [131]. Plotting the reduced dimensions on a two-dimensional plane will visually reveal the clustered ethnic/geographic groups. A reference panel can contain many ethnic/geographic groups, and the number of samples per group can be in the thousands. This mostly depends on the type of data that are available. To ensure the privacy of the samples in the reference panel, the only data that should be used is a unique identifier for each sample that does not reflect a connection to the identity, the genetic data, or the confirmed ethnic/geographic group.

After the reference panel is created, the individual's ethnic/geographic origins can be estimated by comparing the individual's sample to the panel. As SNPs are not inherited individually but in long segments, it is pointless to analyze SNPs as their own dimension. Instead, the best way to compare the samples is by looking for similar combinations of SNPs in certain locations. To do this, all samples have SNPs placed in corresponding groups based on their chromosome/location. There can be around 1000 segmented groups per sample. In addition, each segment group will have two sets of values as DNA is inherited from two sources, mother and father, and these two sources can be from different ethnic/geographic groups. Due to the nature of this data, a Hidden Markov Model (HMM) is used to determine the ethnic/geographic origin for each separate segment. Then, an estimate of the overall percentage of ethnic/geographic make-up can be derived [130]. The significantly represented ethnic/geographic groups can be reported to the individual to inform them of their estimated ethnic/geographic origins.

Genealogy reports give the individual a list of possible relatives based on the genetic distance between the individual's genetic data and other samples in a database, without the use of a reference panel. To measure genetic distance, the unit, centiMorgan, is used. A centiMorgan value of 3300 or higher corresponds to a parent–child relationship, a value between 2200 and 3300 corresponds to a grand parent-child or full sibling relationship, a value between 1300 and 2200 corresponds to an aunt/uncle-child or half-sibling relationship, a value between 650 and 1300 corresponds to great grand parent-child or first cousin relationship, and so on [114]. It is easy to determine if someone has an identical twin or parent–child relationship, but as the centiMorgan value decreases, it becomes more difficult to identify relatives.

To determine the centiMorgan value, identity-by-descent (IBD) segments are used, which are long chromosome segments that two people share. The difficulty in this type of analysis is searching for IBD segments. Many sequencing technologies used in this application do not sequence the entire genome but only sequence a small percentage. In addition, it is difficult to determine from whom the sequence was inherited from the mother or the father. These challenges make it a difficult search for IBD segments. The phasing process is meant to reduce these difficulties.

Once IBD segments are found, only the segments above 7 centiMorgans in length are used to calculate the overall centiMorgan score. The more matching segments there are, the more likely the two samples being compared are related.

6.2 Genotype-based method

6.2.1 HIR match

The definition of half identical region (HIR) match, first appearing in Leon Kull's HIR search project, is that two individuals share a genome where half of its SNP sites are the same for both individuals [132]. Specifically, HIR is a DNA segment where alleles on the region have the characteristic that at least one of the two alleles from one individual's two paired chromosomes in genotype form match at least one of the two alleles from the other individual's two paired chromosomes in genotype form.

Henn et al. propose a genotype-based IBD detection algorithm, where their approach is essentially a pairwise comparison among an individual's unphased genotype data [133]. The way they find the shared IBD segment (also called an IBD half segment) between two individuals is by determining the matched segment boundaries (i.e., two SNP sites) and whether the length of the matched segment between the boundaries meets their criteria. To find an IBD half segment, first, they locate the start position of the segment. They search a *homozygous* SNP site on both individuals' genotypes (i.e., a locus where two alleles from two haplotypes are identical). They consider the locus as a potential start position of the IBD half segment, if the two individuals have opposite homozygotes on the same locus (i.e., one individual has two "A" (or "C") alleles and the other individual has two "G" (or "T") alleles on the same site). Then, they scan through the genotype starting from the potential site, and compare alleles of each site between the two individuals. If at least one of the alleles is identical between two individuals' genotypes, they consider the site to be a match (i.e., an HIR match). The end position of an IBD half segment is defined the same way as the start position, that is, the end position is a *homozygous* SNP site on both individuals' genotypes, and the two individuals have opposite *homozygotes* on that site. At this point, the potential IBD half segment is formed. The criteria the authors use to identify the IBD half segment is that the length of the segment found is at least 5 centiMorgans and it contains at least 400 *homozygous* sites. The authors consider the total number of all found IBD half segments between two individuals to be their shared IBD segments if the longest IBD half segment found between the two individuals is at least 7 centiMorgans. Based on the authors' simulations, they observed IBD segments whose length is at least 7 centiMorgans over 90% of the time. Thus, the criteria resulted in high accuracy of the proposed genotype-based IBD detection algorithm.

6.3 Haplotype-based method

6.3.1 GERMLINE

GERMLINE has been designed to find all IBD segments among all pairs in a given genotype panel [134]. The input files for GERMLINE have phased haplotypes in the PLINK format. GERMLINE searches for seeds (exact matches) of haplotypes between two individuals and then extends the matches if the number of mismatches does not exceed a given threshold. Exact matches are detected efficiently using hash tables. Basically, two exact matches are hashed into the same values and therefore all matches can be extracted efficiently. To achieve accurate results, the data must be well-phased. A mode of the program can extend the seeds from any haplotypes allowing for switch errors. However, it has been shown that switching the haplotype may increase the number of false positives significantly [116].

AncestryDNA developed an extension of GERMLINE, called J-GERMLINE [114], which works with growing databases without recomputing the IBD segments given a new set of queries. The underlying strategy is very similar to GERMLINE where exact matches of haplotypes are found efficiently and then extended if the number of mismatches does not exceed a given threshold.

There are two main issues regarding the efficiency of GERMLINE/J-GERMLINE. Although the exact matches can be found to be linear to the sample size, the total number of matches, especially for short matches, may be quadratic in panels that comprise related individuals. Thus, it requires a large amount of memory to store exact seed matches across the chromosome. Another drawback of using hash tables is the inflexibility of searching for exact matches. Basically, the minimum match should have been determined and preprocessed beforehand.

6.3.2 PBWT-based method, RaPID and PBWT-query

Positional Burrows–Wheeler transform (PBWT) facilitates a fast approach to search for exact matches and also to compress haplotype data [86]. The memory consumption of PBWT is significantly lower than that of hashing methods such as GERMLINE. Furthermore, PBWT can be used to find exact matches of any arbitrary length. The basic idea of PBWT is to sort the haplotype sequences based on their reversed prefix order at each site. By doing so, all matches will be placed adjacent to each other. A data structure called divergence array keeps track of the matches between any haplotype sequence and its immediate neighboring sequence. All matches greater than a given length are separated by a haplotype sequence whose divergence value is larger than $k - L$, where k denotes the current site index and L denotes the minimum length.

Despite the flexibility of PBWT for finding matches of any given length, it cannot tolerate mismatches. Mismatches can occur due to genotyping errors or random mutations. Mutation rates are very low, but the genotyping error rate is currently expected to be 0.1%–0.25%. A simple solution to deal with genotyping error is to search for short seeds and try to extend the matches. However, as for GERMLINE, short seeds are abundant in a real genotype panel. On the other hand, not all the short matches are of interest when searching for relatives of individuals closer to a given degree of relatedness or in the last few generations. Random projection-based IBD detection (RaPID) [135] searches directly for a given target length while tolerating genotyping errors. The basic idea of RaPID is to create subpanels of the original panels. The original panel is divided into windows for a given size and one site is selected within each window. PBWT is then applied to find all exact matches in each subpanel and the results of multiple runs are merged. If two segments are identical, then it is more likely that those matches are reported multiple times.

PBWT data structures enable fast identification of exact matches between all individuals in a panel. Searching for a new query in a panel efficiently cannot be directly solved by PBWT. However, the PBWT data structures can be exploited to search for a new query. PBWT-Query [136] finds the virtual position of the given query at each position in the PBWT panel. All the long matches are stored in the same block in the PBWT arrays; hence, the matches can be enumerated efficiently by scanning up and down in the block of matches. Two more matrices in addition to PPA (positional prefix array) and DA (divergence array) have to be stored. The time complexity is further reduced to the chromosome length or number of matches by introducing additional matrices. The additional precomputed matrices, called LEAP (linked equal/alternating positions) arrays, preclude the unnecessary scanning of the blocks of matches at each site. LEAP arrays consist of eight matrices that contain indices and divergence values. For each value at each site, the indices of alternating and the same alleles are stored, including the maximum divergence values between them. LEAP arrays allow us to jump between the potential matches that are not going to be reported at each site. Note that a match has to be reported if it terminates at each site.

6.3.3 Refined-IBD

Additional information beyond a haplotype match can also be exploited while searching for IBD segments. Refined-IBD [137] first identifies candidate IBD segments using haplotype matches. The candidate segments are regions in which two individuals share a haplotype segment that is greater than a given threshold. For each candidate IBD segment, two likelihoods using haplotype frequencies are computed as follows: (1) the likelihood that one haplotype is a shared IBD, (2) the likelihood of a non-IBD model, where no IBD is shared. Then, the LOD score (log of the likelihood ratio) is computed. Candidate segments with an LOD score greater than a specified threshold are reported as IBD segments.

RefinedIBD relies heavily on the provided panel because haplotype frequency is a major parameter to compute the LOD scores. Basically, the assumption is that a common haplotype is unlikely to be a result of recent IBD, whereas a shared haplotype that is very rare is likely to be IBD. This assumption limits the usage of the algorithm, especially for sequencing data where a large number of sites have a very low minor allele frequency. To solve this issue, sites with low minor allele frequencies have to be filtered out, which ignores some potentially useful information.

Another challenge with RefinedIBD is that the trimming and usage of the candidate IBD segments usually result in an underestimation of the number of IBD segments. Long IBD segments are often reported as a set of short IBD segments with small-to-moderate gaps between them. To solve this issue, the authors suggested merging the results of RefinedIBD, and they also provided a tool to merge the results. RefinedIBD is part of the BEAGLE software. The input file should be provided as VCF format, and if the data is not phased, the program will phase the data first and then search for IBD segments among the phased haplotypes.

6.4 Benchmarking of IBD detection: runtime, power, and accuracy

In run time, RaPID is faster than both GERMLINE and RefinedIBD. For sequencing data, RaPID's runtime is orders of magnitudes faster than GERMLINE. It facilitates an efficient tool for analyzing large cohorts of phased data, such as UK Biobank, without requiring extensive resources.

Simulation results have also shown that RaPID can achieve comparable or higher performance regarding detection power and accuracy for long IBD segments. The accuracy of GERMLINE is comparable or worse than RaPID's while the accuracy of RefinedIBD is higher for short IBD segments. RefinedIBD uses additional information at the cost of increased runtime. In general, RefinedIBD is more accurate for shorter IBD segments, but for longer segments (e.g., 3 cM or above), the differences in accuracy are negligible. The accuracy of RefinedIBD may also decrease due to the merging of the results. Fig. 4.9 shows the benchmarking results of RaPID, RefinedIBD, and GERMLINE in a simulated population with known true IBDs. Power is defined as the average proportion of true IBD segments that have been reported. Accuracy is defined as the number of correctly detected IBDs that share at least 50% overlap with the true IBDs over the number of reported IBDs. Length discrepancy is the root mean squared difference of lengths between the correctly detected IBDs and the true IBDs. In the Power chart of Fig. 4.9, RaPID has higher power compared to other tools while the accuracy of RefinedIBD is higher for shorter IBD segments.

7. Practical methods

7.1 Methods used by DTC companies

DTC companies have similar DNA data processing pipelines to prepare the raw genetic data and perform the matches. In general, all companies use pairwise comparison with a customized genetic distance threshold for an individual's genetic data to find matches from the company's reference population database. The matching method that most DTC companies use is haplotype-based. This means that the intermediate

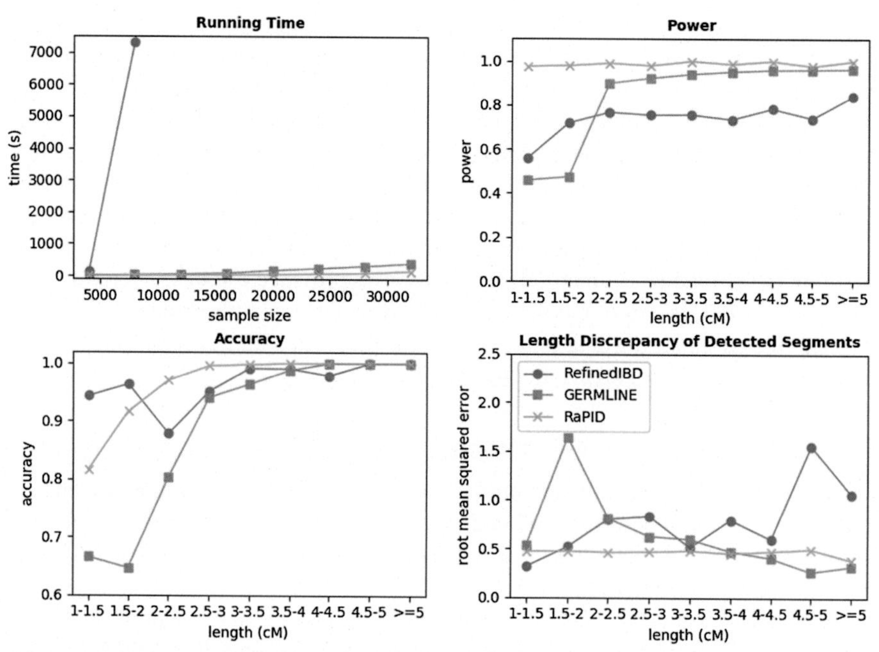

FIGURE 4.9 Benchmarking results of RaPID, GERMLINE and RefinedIBD in a simulated population with known true IBDs [135] (CC BY 4.0 [24]).

genetic data measured by DNA microarray, or genotype, need to be phased into haplotypes. The popular phasing method is a statistical estimation based on the reference population. In most cases, imputation is necessary because there are positions on chromosomes not collected by DNA chip. To increase the accuracy of the inference, most DTC companies use statistical-based imputation methods to fill in those ungenotyped positions with estimated allele values. To speed up the matching process and raise the confidence level of the matches, some DTC companies apply divide and conquer methodology to perform pairwise comparison in parallel (i.e., MapReduce framework), and run statistical classification methods to verify the match results afterward. The methods used by each of the DTC companies can be found in Table 4.4.

8. Challenges and unmet needs

8.1 Ancestry bias

The result of the genealogical search depends on the ethnicities represented by the database. In addition, even if an ethnic group is well represented, the results may depend on the recent genealogical histories of the ethnic group. For example, MyHeritage has some special treatment for Ashkenazi Jews. Also, it is a common practice to use a higher cutoff length for IBD segment searches involving Hispanic people.

Table 4.4 Methods used by each DTC company.

Company	23andMe [112,133]	FamilyTreeDNA [138]	AncestryDNA [114]	MyHeritage [139,140]	GEDmatch
Phasing	Not applicable	Unknown method	Underdog method (a statistical method similar to BEAGLE [40])	Statistical method	Parental Phasing [141], Visual Phasing [142]
Criteria for matching segments	7 cMs and at least 700 SNPs for the first half-identical region; 5 cMs and 700 SNPs for additional half-identical regions and for individuals the customer is sharing with [143]	Option 1: 9 cMs and 500 SNPs for one half-identical region; Option 2: 7.7 cMs for the first half-identical region and a total of at least 20 cMs (including the shorter matching HIRs between 1 cM and 7 cMs); Option 3: 5.5 cMs and at least 500 SNPs for the first half-identical region for about 1% of customers who come from specific non-European populations [143]	6 cMs per segment before the timber algorithm is applied and a total of at least 6 cMs after timber is applied [143]	8 cM for the first matching segment and at least 6 cMs for the 2nd matching segment; 12 cM for the first matching segment in people whose ancestry is at least 50% Ashkenazi Jewish [143]	Pairwise comparison of diploid genotype data by using HIR match with user-defined cM threshold [144]
Verification	Not applicable	Not applicable	Scoring detected IBD matches by using the timber algorithm to adjust their genetic distances (de-emphasize the detected IBD segment if the SNP window has a large number of times it overlaps the IBD segment)	Statistical classification method to reject false positives and determine the confidence level of the matches (high, medium, or low)	The user-defined "confidence" threshold. Combines the matching algorithm's "Q" score and the user-defined "Precision" score to filter matches [144]

8.2 Phasing imperfection

If phasing information is accurately available, haplotype-based methods are obviously advantageous. However, current phasing is not necessarily accurate enough, so precautions are often taken in practice. As phasing quality improves in the future, haplotype-based methods will be the mainstay.

8.3 Benchmarking of genealogical search

Currently, there is no standard benchmarking available. Establishing commonly accessible datasets and commonly agreed sets of benchmarking protocols will help bring this actively developing field into maturation.

9. Privacy concerns

Privacy concerning personal genomics is an issue that has slowly evolved since the conclusion of the Human Genome Project (HGP) in 2003.

Shortly after the completion of the HGP, publicly accessible databases, like HapMap and 1000 Genomes Project, were the infrastructure implemented to facilitate the sharing of genomic data. The ability to freely share this data allowed for the growth of knowledge in the personal genomics field [145]. Shortly after the system of data sharing was developed, privacy concerns began to take shape as the ability to identify individuals from anonymous data became easier with the aid of advancements in computing. Researchers showed many ways that open-source material could be used to identify previously anonymous individuals. For example, Nils Homer presented a method using high-density single nucleotide polymorphism (SNP) genotyping microarrays to assert whether an individual was in a dataset [146]. Security deficits in some public databases, like HapMap, led to them being shut down [4].

To limit data exposure, the research community shifted to a controlled-access model. Implemented using databases called biobanks, it allowed for regulation and monitoring of data flow [147]. Restricting access to trusted parties worked effectively in preventing malicious uses of the data in research databases.

Treating personal genomic data like any other sensitive medical data is not enough to protect the privacy of individuals and their families. Genomic data hold unique identifiers, information on ancestry, and can determine an individual's susceptibility to certain health issues. To many, the exposure of this information is concerning. The fear of genetic discrimination is a rational thought to anyone, and for that reason countries like the United States have put measures in place to protect their citizens. The Genetic Information Nondiscrimination Act of 2008 (GINA) was introduced by the US federal government to prevent employers and health insurance agencies from using personal genetic information to make decisions. The goal of this act was to give a sense of protection to individuals and make them feel more comfortable in completing genetic testing as part of their normal medical care [148].

Along with GINA, the Standards for Privacy of Individually Identifiable Health Information in the Health Insurance Portability and Accountability Act of 1996 (HIPAA) regulate the privacy of a patient's personal health information. Organizations that must adhere to this law include health providers, health insurers, the companies holding onto the health data, and all those working with these organizations [147].

Beyond the clinical and research realm, the one area where privacy concerns of personal genomics run high is with the evidence collection process in judicial systems.

In recent years, the growth in popularity of DTC genetic testing companies has led many individuals to submit their DNA to receive results regarding their genealogy and ancestry. Most of these companies allow their customers to download their genetic data, which includes around a quarter-million SNPs [149]. With this raw data, third-party services, like GEDmatch, have formed to allow individuals to submit their genetic data and find relatives who have done the same [150]. As a result, these third-party services have amassed large databases of personal genetic information. GEDmatch has over 1 million people as of 2018, which has been used to help identify suspects in serious criminal investigations [27]. In 2018, Erlich et al. predicted that only 2% of a target population would need to have their DNA in a database to show at least one third cousin match for everyone in the target population [149]. The implications of this prediction show how easy it could be for a law enforcement investigation to narrow down a criminal suspect list by comparing unidentified DNA evidence to a large database like that of GEDmatch. Most concerns would come from the individuals put on these criminal suspect lists because they have a familial genetic relationship with an individual who submitted their genetic data to a third-party service website.

In addition, in 2018, a survey of 1500 people was conducted to better understand the public opinion on law enforcement agencies using genetic genealogy websites and DTC databases to conduct criminal investigations. The results showed that most people support this method of investigation as long as the crime being investigated was violent in nature, involving missing persons, or involving a crime against a child. The people surveyed did not support this method of investigation to find suspects of nonviolent crimes [151].

Still, with public opinion in favor of the "familial searching" method for certain criminal investigations, there are people who feel the justice for the victim overshadows the rights of the citizen. Some believe the courts and prosecutors could incorrectly come to conclusions from improperly presented genetic evidence. The fact that certain genetic material was found at a crime scene does not implicate the person, just that the person was at the location at some point in time. In addition, the evidence collection process could have been tainted by an unrecognized mistake. All of these factors need to be considered to properly protect the rights of the citizen [152]. For example, a court case involving a 2016 homicide in Texas was recently thrown out by a judge because the protocols used to analyze the DNA evidence were not carried out properly [153]. This issue just stresses the importance of properly trained individuals when the result of their work could determine the outcome of high profile court proceedings.

10. Conclusions

In this chapter, we discussed historic trends and recent developments of genetic genealogical search. With tens of millions of consumers with whole-genome genetic profiles at hand, the power of genetic genealogical search is truly unleashed. We are fortunate enough to witness such power and how society is adjusted to it. With the understanding of the technical aspects detailed earlier, we hope the reader is well-informed and will be able to help lead the world to a brighter future.

Acknowledgments

This work was supported by the US National Institutes of Health [R01HG010086].

References

[1] About the international HapMap project. June 4, 2012 [cited 9 Oct 2019]. Available from: https://www.genome.gov/11511175/about-the-international-hapmap-project-fact-sheet.

[2] The International HapMap Consortium. The international HapMap project. Nature 2003:789−96. https://doi.org/10.1038/nature02168.

[3] International HapMap Consortium. Integrating ethics and science in the international HapMap project. Nature Reviews Genetics 2004;5:467−75.

[4] NCBI retiring HapMap resource. June 16, 2016 [cited 10 Sep 2019]. Available from: https://www.ncbi.nlm.nih.gov/variation/news/NCBI_retiring_HapMap/.

[5] Meeting report: a workshop to plan a deep catalog of human genetic variation. September 17, 2007 [cited 11 Sep 2019]. Available from: https://www.internationalgenome.org/sites/1000genomes.org/files/docs/1000Genomes-MeetingReport.pdf.

[6] About IGSR and the 1000 genome project. [cited 11 Sep 2019]. Available from: https://www.internationalgenome.org/about#1000G_PROJECT.

[7] 1000 Genomes Project Consortium, Abecasis GR, Altshuler D, Auton A, Brooks LD, Durbin RM, et al. A map of human genome variation from population-scale sequencing. Nature 2010;467:1061−73.

[8] Devuyst O. The 1000 genomes project: welcome to a new world. Peritoneal Dialysis International 2015;35:676−7.

[9] UK biobank. About UK biobank. February 7, 2019 [cited 18 Sep 2019]. Available from: https://www.ukbiobank.ac.uk/about-biobank-uk/.

[10] UK Biobank. Researchers. June 19, 2019 [cited 18 Sep 2019]. Available from: https://www.ukbiobank.ac.uk/researchers/.

[11] UK Biobank. Approved research summary. February 21, 2019 [cited 18 Sep 2019]. Available from: http://www.ukbiobank.ac.uk/approved-research/.

[12] Sarata AK. Genetic ancestry testing. 2008.

[13] Hamilton A. Invention of the year: the retail DNA test. October 29, 2008 [cited 19 Sep 2019]. Available from: http://content.time.com/time/specials/packages/article/0,28804,1852747_1854493,00.html.

[14] 23andMe. About us. [cited 19 Sep 2019]. Available from: https://mediacenter.23andme.com/company/about-us/.
[15] Wojcicki A. FDA warning letter to 23andMe, Inc. November 22, 2013 [cited 19 Sep 2019]. Available from: https://perma.cc/9GRR-8WCT.
[16] Rabbani B, Tekin M, Mahdieh N. The promise of whole-exome sequencing in medical genetics. Journal of Human Genetics 2014:5–15. https://doi.org/10.1038/jhg.2013.114.
[17] McAllister BF. Growth in DTC genetic testing. June 30, 2015 [cited 19 Sep 2019]. Available from: https://wiki.uiowa.edu/display/2360159/2015/06/26/Growth+in+DTC+Genetic+Testing.
[18] AncestryDNA. Our story. [cited 19 Sep 2019]. Available from: https://www.ancestry.com/corporate/about-ancestry/our-story.
[19] Check Hayden E. The rise and fall and rise again of 23andMe. Nature 2017;550:174–7.
[20] FDA allows marketing of first direct-to-consumer tests that provide genetic risk information for certain conditions. April 6, 2017 [cited 20 Sep 2019]. Available from: https://www.fda.gov/news-events/press-announcements/fda-allows-marketing-first-direct-consumer-tests-provide-genetic-risk-information-certain-conditions.
[21] 23andMe granted first FDA authorization for direct-to-consumer genetic test on cancer risk. March 6, 2018 [cited 20 Sep 2019]. Available from: https://mediacenter.23andme.com/press-releases/23andme-granted-first-fda-authorization-direct-consumer-genetic-test-cancer-risk/.
[22] Regalado A. 2017 was the year consumer DNA testing blew up. February 12, 2018 [cited 20 Sep 2019]. Available from: https://www.technologyreview.com/s/610233/2017-was-the-year-consumer-dna-testing-blew-up/.
[23] Khan R, Mittelman D. Consumer genomics will change your life, whether you get tested or not. Genome Biology 2018;19:120.
[24] Creative Commons. Attribution 4.0 International (CC BY 4.0). Available from: https://creativecommons.org/licenses/by/4.0/.
[25] Larkin L. DNA tests. In: The DNA Geek; September 18, 2019 [cited 20 Sep 2019]. Available from: https://thednageek.com/dna-tests/.
[26] Ram N, Guerrini CJ, McGuire AL. Genealogy databases and the future of criminal investigation. Science 2018;360:1078–9.
[27] Kennett D. Using genetic genealogy databases in missing persons cases and to develop suspect leads in violent crimes. Forensic Science International 2019;301:107–17.
[28] DNA.land. Frequently asked questions. Purpose. [cited 20 Sep 2019]. Available from: https://dna.land/faq#purpose.
[29] Johannsen W. Arvelighedslærens elementer: forelæsninger holdte ved Københavns universitet. 1905.
[30] Noble D. Genes and causation. Philosophical Transactions of the Royal Society A: Mathematical, Physical and Engineering Sciences 2008:3001–15. https://doi.org/10.1098/rsta.2008.0086.
[31] Watson JD, Crick FHC. Molecular structure of nucleic acids: a structure for deoxyribose nucleic acid (Reprinted from Nature, April 25, 1953) Nature 1969:470–1. https://doi.org/10.1038/224470a0.
[32] Collins FS, Brooks LD, Chakravarti A. A DNA polymorphism discovery resource for research on human genetic variation. Genome Research 1998;8:1229–31.

[33] Wood EJ. The encyclopedia of molecular biology. Biochemical Education 1995:105. https://doi.org/10.1016/0307-4412(95)90659-2.

[34] Lodish HF, Berk, Lodish, zipursky, Matsudaria, Berk A, Darnell JE, Zipursky SL, Baltimore D, et al. Molecular cell biology. Scientific American Library; 2000.

[35] Moore PD, Barry Cox C, Ladle RJ. Biogeography: an ecological and evolutionary approach. 10th ed. Wiley-Blackwell; 2018.

[36] Austin CP. Talking glossary of genetic terms. Haploid. In: National Human Genome Research Institute. [cited 19 Aug 2019]. Available from: https://www.genome.gov/genetics-glossary/haploid.

[37] Carr SM, Dawn Marshall H, Duggan AT, Flynn SMC, Johnstone KA, Pope AM, et al. Phylogeographic genomics of mitochondrial DNA: highly-resolved patterns of intra-specific evolution and a multi-species, microarray-based DNA sequencing strategy for biodiversity studies. Comparative Biochemistry and Physiology Part D: Genomics and Proteomics 2008:1–11. https://doi.org/10.1016/j.cbd.2006.12.005.

[38] The 1000 Genomes Project Consortium. A global reference for human genetic variation. Nature 2015:68–74. https://doi.org/10.1038/nature15393.

[39] Stephens M, Smith NJ, Donnelly P. A new statistical method for haplotype reconstruction from population data. The American Journal of Human Genetics 2001;68:978–89.

[40] Browning SR, Browning BL. Rapid and accurate haplotype phasing and missing-data inference for whole-genome association studies by use of localized haplotype clustering. The American Journal of Human Genetics 2007;81:1084–97.

[41] Howie BN, Donnelly P, Marchini J. A flexible and accurate genotype imputation method for the next generation of genome-wide association studies. PLoS Genetics 2009;5:e1000529.

[42] Li Y, Willer CJ, Ding J, Scheet P, Abecasis GR. MaCH: using sequence and genotype data to estimate haplotypes and unobserved genotypes. Genetic Epidemiology 2010;34:816–34.

[43] Delaneau O, Marchini J, Zagury J-F. A linear complexity phasing method for thousands of genomes. Nature Methods 2011;9:179–81.

[44] Delaneau O, Zagury J-F, Marchini J. Improved whole-chromosome phasing for disease and population genetic studies. Nature Methods 2013;10:5–6.

[45] Loh P-R, Palamara PF, Price AL. Fast and accurate long-range phasing in a UK Biobank cohort. Nature Genetics 2016;48:811–6.

[46] Loh P-R, Danecek P, Palamara PF, Fuchsberger C, Reshef YA, Finucane HK, et al. Reference-based phasing using the haplotype reference Consortium panel. Nature Genetics 2016;48:1443–8.

[47] Andrews CA. The hardy-weinberg principle. In: Nature education knowledge; 2010 [cited 22 Aug 2019]. Available from: https://www.nature.com/scitable/knowledge/library/the-hardy-weinberg-principle-13235724/.

[48] What are genome-wide association studies. In: Genetics home reference; September 10, 2019 [cited 12 Sep 2019]. Available from: https://ghr.nlm.nih.gov/primer/genomicresearch/gwastudies.

[49] Encoding structural variants in VCF (Variant Call Format) version 4.0. In: IGSR: The International Genome Sample Resource. [cited 23 Aug 2019]. Available from: https://www.internationalgenome.org/wiki/Analysis/Variant%20Call%20Format/VCF%20(Variant%20Call%20Format)%20version%204.0/encoding-structural-variants.

[50] Danecek P, Auton A, Abecasis G, Albers CA, Banks E, DePristo MA, et al. The variant call format and VCFtools. Bioinformatics 2011:2156–8. https://doi.org/10.1093/bioinformatics/btr330.
[51] Global alliance for genomics and health. [cited 23 Aug 2019]. Available from: https://www.ga4gh.org/.
[52] Samtools. The variant call format (VCF) version 4.2 specification. July 8, 2019 [cited 23 Aug 2019]. Available from: https://samtools.github.io/hts-specs/VCFv4.2.pdf.
[53] IGSR: The International Genome Sample Resource. [cited 23 Aug 2019]. Available from: https://www.internationalgenome.org/.
[54] ExAC: Exome Aggregation Consortium. [cited 23 Aug 2019]. Available from: http://exac.broadinstitute.org/.
[55] GATK: Genome Analysis Toolkit. [cited 23 Aug 2019]. Available from: https://software.broadinstitute.org/gatk/.
[56] VEP: Variant Effect Predictor. [cited 23 Aug 2019]. Available from: https://useast.ensembl.org/info/docs/tools/vep/index.html.
[57] EVA: European Variation Archive. [cited 23 Aug 2019]. Available from: https://www.ebi.ac.uk/eva/.
[58] dbSNP: Single Nucleotide Polymorphism Database. [cited 23 Aug 2019]. Available from: https://www.ncbi.nlm.nih.gov/snp/.
[59] . UK10K: 10,000 UK Genome Sequences Consortium. [cited 23 Aug 2019]. Available from: https://www.uk10k.org/.
[60] NHLBI GO ESP: NHLBI GO Exome Sequencing Project. [cited 23 Aug 2019]. Available from: https://evs.gs.washington.edu/EVS/.
[61] Human genome overview. In: Genome reference consortium; February 27, 2009 [cited 26 Aug 2019]. Available from: https://www.ncbi.nlm.nih.gov/grc/human.
[62] Raw data technical details. In: 23andMe. [cited 26 Aug 2019]. Available from: https://customercare.23andme.com/hc/en-us/articles/115004459928-Raw-Data-Technical-Details.
[63] Downloading AncestryDNA raw data. In: AncestryDNA. [cited 26 Aug 2019]. Available from: https://support.ancestry.com/s/article/Downloading-AncestryDNA-Raw-Data.
[64] How should I interpret my raw DNA data. In: MyHeritage. [cited 26 Aug 2019]. Available from: https://faq.myheritage.com/en/article/how-should-i-interpret-my-raw-dna-data.
[65] How do I read my Family Finder raw data file. In: FamilyTreeDNA. [cited 26 Aug 2019]. Available from: https://www.familytreedna.com/learn/autosomal-ancestry/universal-dna-matching/read-family-finder-raw-data-file/.
[66] Purcell S, Neale B, Todd-Brown K, Thomas L, Ferreira MAR, Bender D, et al. PLINK: a tool set for whole-genome association and population-based linkage analyses. The American Journal of Human Genetics 2007;81:559.
[67] Chang CC, Chow CC, Tellier LC, Vattikuti S, Purcell SM, Lee JJ. Second-generation PLINK: rising to the challenge of larger and richer datasets. Gigascience 2015;4. https://doi.org/10.1186/s13742-015-0047-8.
[68] PLINK: basic usage and data formats. January 25, 2017 [cited 28 Aug 2019]. Available from: http://zzz.bwh.harvard.edu/plink/data.shtml.
[69] PLINK: BED file format. January 25, 2017 [cited 28 Aug 2019]. Available from: http://zzz.bwh.harvard.edu/plink/binary.shtml.

[70] Band G, Marchini J. BGEN: a binary file format for imputed genotype and haplotype data. bioRxiv 2018:308296. https://doi.org/10.1101/308296.
[71] UK Biobank. [cited 28 Aug 2019]. Available from: https://www.ukbiobank.ac.uk/.
[72] Bycroft C, Freeman C, Petkova D, Band G, Elliott LT, Sharp K, et al. The UK Biobank resource with deep phenotyping and genomic data. Nature 2018;562:203–9.
[73] Wellcome Sanger Institute. Wellcome Trust Case Control Consortium. [cited 30 Aug 2019]. Available from: https://www.wtccc.org.uk/ccc2/.
[74] Malaria Genomic Epidemiology Network (GEN). [cited 30 Aug 2019]. Available from: https://www.malariagen.net/human.
[75] Avon Longitudinal Study of Parents and Children (ALSPAC). In: University of Bristol. [cited 30 Aug 2019]. Available from: http://www.bristol.ac.uk/alspac/.
[76] Gavin Band JM. The BGEN format. [cited 30 Aug 2019]. Available from: https://www.well.ox.ac.uk/~gav/bgen_format/.
[77] Gavin Band JM. The BGEN format: Overview. [cited 30 Aug 2019]. Available from: https://www.well.ox.ac.uk/~gav/bgen_format/spec/latest.html.
[78] Band G. BGEN: wiki. In: Bitbucket; July 15, 2017 [cited 30 Aug 2019]. Available from: https://bitbucket.org/gavinband/bgen/wiki/bgenix.
[79] Description of the UK Biobank genotype and imputed data sets. July 2017 [cited 30 Aug 2019]. Available from: http://www.ukbiobank.ac.uk/wp-content/uploads/2017/07/ukb_genetic_file_description.txt.
[80] Zheng X, Levine D, Shen J, Gogarten SM, Laurie C, Weir BS. A high-performance computing toolset for relatedness and principal component analysis of SNP data. Bioinformatics 2012;28:3326–8.
[81] Zheng X, Gogarten SM, Lawrence M, Stilp A, Conomos MP, Weir BS, et al. SeqArray—a storage-efficient high-performance data format for WGS variant calls. Bioinformatics 2017;33:2251–7.
[82] R Interface to CoreArray genomic data structure (GDS) files. In: Bioconductor. [cited 30 Aug 2019]. Available from: http://www.bioconductor.org/packages/release/bioc/html/gdsfmt.html.
[83] Data management of large-scale whole-genome sequence variant calls. In: Bioconductor. [cited 30 Aug 2019]. Available from: http://bioconductor.org/packages/release/bioc/html/SeqArray.html.
[84] TOPMed. In: National heart, lung and blood institute trans-omics for precision medicine. [cited 30 Aug 2019]. Available from: https://www.nhlbiwgs.org/.
[85] Gogarten SM, Sofer T, Chen H, Yu C, Brody JA, Thornton TA, et al. Genetic association testing using the GENESIS R/Bioconductor package. Bioinformatics 2019. https://doi.org/10.1093/bioinformatics/btz567.
[86] Durbin R. Efficient haplotype matching and storage using the positional Burrows–Wheeler transform (PBWT). Bioinformatics 2014;30:1266–72.
[87] Slatkin M. Linkage disequilibrium — understanding the evolutionary past and mapping the medical future. Nature Reviews Genetics 2008;9:477.
[88] Li H. BGT: efficient and flexible genotype query across many samples. Bioinformatics 2016;32:590.
[89] Tsukiyama T, Kondo Y, Kakuse K, Saba S, Ozaki S, Itoh K. Method and system for data compression and restoration. US Patent. 4586027. 1986. Available from: https://patentimages.storage.googleapis.com/53/aa/40/a8c81c61d4eaac/US4586027.pdf.
[90] Li H. BGT. In: github. [cited 31 Aug 2019]. Available from: https://github.com/lh3/bgt.

[91] Kelleher J, Wong Y, Wohns AW, Fadil C, Albers PK, McVean G. Inferring whole-genome histories in large population datasets. Nature Genetics 2019;51:1330–8.

[92] Nei M. Molecular evolutionary genetics. 1987. https://doi.org/10.7312/nei-92038.

[93] Genetic distance. In: FamilyTreeDNA. [cited 4 Sep 2019]. Available from: https://www.familytreedna.com/learn/faq-items/genetic-distance/.

[94] Griffiths AJF, Miller JH, Suzuki DT, Lewontin RC, Gelbart WM. Linkage maps. An introduction to genetic analysis. 7th ed. W. H. Freeman; 2000.

[95] Lobo ISK. Thomas hunt morgan, genetic recombination, and gene mapping. In: Nature education; 2008 [cited 4 Sep 2019]. Available from: https://www.nature.com/scitable/topicpage/thomas-hunt-morgan-genetic-recombination-and-gene-496/.

[96] Lodish H, Berk A, Matsudaira P, Kaiser CA. Molecular cell biology 5th edition, modern genetic analysis 2nd edition & Cd-rom. W H Freeman & Company; 2004.

[97] centiMorgan. In: FamilyTreeDNA. [cited 4 Sep 2019]. Available from: https://www.familytreedna.com/learn/faq-items/centimorgan-cm/.

[98] centiMorgans. In: YourDNAGuide. [cited 4 Sep 2019]. Available from: https://www.yourdnaguide.com/ydgblog/2019/2/15/centimorgans.

[99] Malécot G. Les Mathématiques de l'hérédité. Masson; 1948.

[100] Nagylaki T. Gustave Malécot and the transition from classical to modern population genetics. Genetics 1989;122:253–68.

[101] Kingman JFC. Origins of the coalescent: 1974–1982. Genetics 2000;156:1461–3.

[102] A second generation human haplotype map of over 3.1 million SNPs. Nature 2007; 449:851–61.

[103] Powell JE, Visscher PM, Goddard ME. Reconciling the analysis of IBD and IBS in complex trait studies. Nature Reviews Genetics 2010;11:800–5.

[104] Thompson EA. Identity by descent: variation in meiosis, across genomes, and in populations. Genetics 2013;194:301.

[105] Identical by descent. In: ISOGG. [cited 5 Sep 2019]. Available from: https://isogg.org/wiki/Identical_by_descent.

[106] Identical by state. In: ISOGG. [cited 5 Sep 2019]. Available from: https://isogg.org/wiki/Identical_by_state.

[107] Bettinger B. The Effect of phasing on reducing false distant matches (or, phasing a parent using GEDmatch). In: The genetic genealogist; July 26, 2017 [cited 5 Sep 2019]. Available from: https://thegeneticgenealogist.com/2017/07/26/the-effect-of-phasing-on-reducing-false-distant-matches-or-phasing-a-parent-using-gedmatch/.

[108] Concepts – identical by descent, state, population and chance. In: DNAeXplained; March 10, 2016 [cited 5 Sep 2019]. Available from: https://dna-explained.com/2016/03/10/concepts-identical-bydescent-state-population-and-chance/.

[109] Gklambauer. Identity by descent. In: Wikipedia; January 23, 2013 [cited 23 Sep 2019]. Available from: https://en.wikipedia.org/wiki/Identity_by_descent.

[110] Creative Commons. Attribution-ShareAlike 3.0 Unported (CC BY-SA 3.0). Available from: https://creativecommons.org/licenses/by-sa/3.0/.

[111] Speed D, Balding DJ. Relatedness in the post-genomic era: is it still useful? Nature Reviews Genetics 2014;16:33–44.

[112] DNA relatives: detecting relatives and predicting relationships. In: 23andMe. [cited 7 Sep 2019]. Available from: https://customercare.23andme.com/hc/en-us/articles/212170958-DNA-Relatives-Detecting-Relatives-and-Predicting-Relationships.

[113] MyHeritage Overhauls their matching algorithm. In: The DNA Geek; January 12, 2018 [cited 7 Sep 2019]. Available from: https://thednageek.com/myheritage-overhauls-their-matching-algorithm/.

[114] Ball CA, Barber MJ, Byrnes J, Carbonetto P, Chahine KG, Curtis RE, Granka JM, Han E, Hong EL, Kermany AR, Myres NM, Keith N, Qi J, Rand K, Wang Y, Willmore L. AncestryDNA matching white paper. In: AncestryDNA; March 31, 2016 [cited 7 Sep 2019]. Available: https://www.ancestry.com/dna/resource/whitePaper/AncestryDNA-Matching-White-Paper.pdf.

[115] Bettinger B. Family tree DNA updates matching thresholds. In: The genetic genealogist; May 24, 2016 [cited 7 Sep 2019]. Available from: https://thegeneticgenealogist.com/2016/05/24/family-tree-dna-updates-matching-thresholds/.

[116] Durand EY, Eriksson N, McLean CY. Reducing pervasive false-positive identical-by-descent segments detected by large-scale pedigree analysis. Molecular Biology and Evolution 2014;31:2212.

[117] Eastman D. Is there any such thing as a half-cousin?. In: Eastman's online genealogy newsletter; January 21, 2015 [cited 10 Sep 2019]. Available from: https://blog.eogn.com/2015/01/21/is-there-any-such-thing-as-a-half-cousin/.

[118] Dictionary BL. What is cousin?. In: The law dictionary; 1910 [cited 10 Sep 2019]. Available from: https://thelawdictionary.org/cousin/.

[119] Degrees of cousin-ness. In: AncestryDNA; March 29, 2014 [cited 10 Sep 2019]. Available from: https://blogs.ancestry.com/ancestry/2014/03/29/degrees-of-cousin-ness/.

[120] Kinship terminology explained (or how to know what to call distant relatives). In: Find my past; June 6, 2016 [cited 10 Sep 2019]. Available from: https://www.findmypast.co.uk/blog/help/kinship-terminology-how-we-refer-to-our-family-relationships.

[121] Creative Commons. Attribution-NoDerivs 2.5 Generic (CC BY-ND 2.5). Available from: https://creativecommons.org/licenses/by-nd/2.5/.

[122] Miko I. Gregor Mendel and the principles of inheritance. In: Nature education; 2008 [cited 12 Sep 2019]. Available from: https://www.nature.com/scitable/topicpage/gregor-mendel-and-the-principles-of-inheritance-593/.

[123] Freeman S, Quillin K, Allison LA, Podgorski G, Black M, Taylor E. Biological science. Pearson; 2019.

[124] Libiger O, Schork NJ. A simulation-based analysis of chromosome segment sharing among a group of arbitrarily related individuals. European Journal of Human Genetics 2007;15:1260–8.

[125] Autosomal DNA statistics. In: ISOGG. [cited 12 Sep 2019]. Available from: https://isogg.org/wiki/Autosomal_DNA_statistics.

[126] Bettinger B. August 2017 update to the shared cM project. In: The genetic genealogist; August 26, 2017 [cited 16 Sep 2019]. Available from: https://thegeneticgenealogist.com/2017/08/26/august-2017-update-to-the-shared-cm-project/.

[127] Mercedes. Beginner's guide to shared centimorgans. In: Who are you made of; February 21, 2018 [cited 16 Sep 2019]. Available from: https://whoareyoumadeof.com/blog/2018/02/21/beginners-guide-shared-centimorgans/.

[128] Donnelly KP. The probability that related individuals share some section of genome identical by descent. Theoretical Population Biology 1983;23:34–63.

[129] Cousin statistics. In: ISOGG. [cited 16 Sep 2019]. Available from: https://isogg.org/wiki/Cousin_statistics.

[130] Ethnicity estimate 2018 white paper. In: AncestryDNA; September 11, 2018 [cited 8 Aug 2019]. Available from: https://www.ancestrycdn.com/dna/static/images/ethnicity/help/WhitePaper_Final_091118dbs.pdf.

[131] Jackson JE. A user's guide to principal components. John Wiley & Sons; 2005.

[132] Half-identical region. In: ISOGG; June 2, 2017 [cited 18 Sep 2019]. Available from: https://isogg.org/wiki/Half-identical_region.

[133] Henn BM, Hon L, Michael Macpherson J, Eriksson N, Saxonov S, Pe'er I, et al. Cryptic distant relatives are common in both isolated and cosmopolitan genetic samples. PLoS One 2012;7. https://doi.org/10.1371/journal.pone.0034267.

[134] Gusev A, Lowe JK, Stoffel M, Daly MJ, Altshuler D, Breslow JL, et al. Whole population, genome-wide mapping of hidden relatedness. Genome Research 2009;19: 318.

[135] Naseri A, Liu X, Tang K, Zhang S, Zhi D. RaPID: ultra-fast, powerful, and accurate detection of segments identical by descent (IBD) in biobank-scale cohorts. Genome Biology 2019;20:1–15.

[136] Naseri A, Holzhauser E, Zhi D, Zhang S. Efficient haplotype matching between a query and a panel for genealogical search. Bioinformatics 2019;35:i233–41.

[137] Browning BL, Browning SR. Improving the accuracy and efficiency of identity-by-descent detection in population data. Genetics 2013;194:459–71.

[138] Family finder – family matching system. In: FamilyTreeDNA. [cited 20 Sep 2019]. Available from: https://www.familytreedna.com/learn/ftdna/ftdna-family-matching-system/.

[139] MyHeritage LIVE conference day 2 – the science behind DNA matching. In: DNAeXplained; November 6, 2018 [cited 20 Sep 2019]. Available from: https://dna-explained.com/2018/11/06/myheritage-live-conference-day-2-the-science-behind-dna-matching/.

[140] Major updates and improvements to MyHeritage DNA matching. In: MyHeritage; January 11, 2018 [cited 20 Sep 2019]. Available from: https://blog.myheritage.com/2018/01/major-updates-and-improvements-to-myheritage-dna-matching/.

[141] Bettinger B. GEDmatch.com adds phasing tool. In: The genetic genealogist; June 7, 2012 [cited 20 Sep 2019]. Available from: https://thegeneticgenealogist.com/2012/06/07/gedmatch-com-adds-phasing-tool/.

[142] Bettinger B. Visual phasing: an example. In: The genetic genealogist; November 21, 2016 [cited 20 Sep 2019]. Available from: https://thegeneticgenealogist.com/2016/11/21/visual-phasing-an-example-part-1-of-5/.

[143] Autosomal DNA testing comparison chart. In: ISOGG. [cited 5 Sep 2019]. Available from: https://isogg.org/wiki/Autosomal_DNA_testing_comparison_chart.

[144] Q matching. In: GEDmatch. [cited 20 Sep 2019]. Available from: https://www.gedmatch.com/Documents/Qdocs.pdf.

[145] Greenbaum D, Sboner A, Mu XJ, Gerstein M. Genomics and privacy: implications of the new reality of closed data for the field. PLoS Computational Biology 2011;7: e1002278.

[146] Homer N, Szelinger S, Redman M, Duggan D, Tembe W, Muehling J, et al. Resolving individuals contributing trace amounts of DNA to highly complex mixtures using high-density SNP genotyping microarrays. PLoS Genetics 2008;4:e1000167.

[147] Shi X, Wu X. An overview of human genetic privacy. Annals of the New York Academy of Sciences 2017;1387:61–72.

[148] Hampton T. Congress passes bill to ban discrimination based on individuals' genetic makeup. Journal of the American Medical Association 2008;299:2493.
[149] Erlich Y, Shor T, Pe'er I, Carmi S. Identity inference of genomic data using long-range familial searches. Science 2018;362:690–4.
[150] Wang C, Cahill TJ, Parlato A, Wertz B, Zhong Q, Cunningham TN, et al. Consumer use and response to online third-party raw DNA interpretation services. Molecular Genetics and Genomic Medicine 2018;6:35–43.
[151] Guerrini CJ, Robinson JO, Petersen D, McGuire AL. Should police have access to genetic genealogy databases? Capturing the Golden State Killer and other criminals using a controversial new forensic technique. PLoS Biology 2018;16:e2006906.
[152] Berkman BE, Miller WK, Grady C. Is it ethical to use genealogy data to solve crimes? Annals of Internal Medicine 2018;169:333–4.
[153] Augenstein S. DNA mixture analysis thrown out of Texas murder trial - but software debate remains. Forensic Magazine June 27, 2018 [cited 17 Sep 2019]. Available from: https://www.forensicmag.com/news/2018/06/dna-mixture-analysis-thrown-out-texas-murder-trial-software-debate-remains.

SECTION

Privacy-preserving techniques for responsible genomic data sharing

CHAPTER

Homomorphic encryption 5

Kim Laine, PhD
Microsoft Research, Redmond WA, United States

1. Overview

A fundamental challenge in storing and processing of biomedical data is the high level of data privacy that handling of such data necessitates. These requirements are sometimes enforced by governments and sometimes stem from liability concerns or fear of negative publicity from data breaches. While public cloud services could in principle provide the needed storage and computation infrastructure, private patient data should not be exposed to the cloud service operators. A simple partial remedy is for the hospital to store only encrypted data in the cloud, but this significantly weakens the capabilities of the service: any computation or operation on the data requires it to be downloaded back to the hospital's system, to be decrypted, and then computed on. Another common solution is to simply give the cloud service access to the decryption key and trust that they implement appropriate access policies to prevent anyone from decrypting all of the data and downloading or sharing it.

This is one of the reasons why hospitals and medical clinics prefer to store their data on premises, rather than outsourcing storage and computation to much more cost-efficient and flexible public cloud services. However, it is extremely costly and challenging even for large hospitals with well-funded IT departments to operate their own data centers securely and reliably. Furthermore, if these systems fail to provide an appropriate level of functionality or if the implemented security measures are seen as inconvenient by, e.g., the clinical staff, it may be tempting for them to instead use personal laptops or tablets as more reliable and convenient alternatives for short-term data storage. Unfortunately, these devices regularly get lost or stolen: the number of health-care-related data breaches has skyrocketed in recent years, exposing potentially private data of over 50% of the entire population of the United States [1,2].

In this chapter we discuss a relatively new cryptographic technology, *homomorphic encryption*, that may solve some of these issues. At a high level, homomorphic encryption allows hospitals to encrypt their data and store it in a public cloud, while at the same time allowing the cloud to compute and operate directly on the encrypted data. These encrypted computations yield encrypted results, which can only be decrypted with a secret key held by the hospital and not shared with the cloud.

Responsible Genomic Data Sharing. https://doi.org/10.1016/B978-0-12-816197-5.00005-X

1.1 Early ideas

In a seminal 1978 paper Rivest et al. [3] discussed cryptographic solutions to data privacy in outsourced storage and computation settings. Recently the rise of massive cloud storage has made these privacy issues more pressing than ever, and creates an unfortunate clash between privacy, functionality, and practicality. As a suggested solution, Rivest et al. presented the concept of *privacy homomorphisms*: special functions that the data bank—the *cloud* in modern terminology—can apply directly on encrypted customer data without requiring access to a secret decryption key. Crucially, such evaluations should only output encrypted results, thus revealing nothing about the input or the output to the cloud. Several examples of privacy homomorphisms were presented and discussed in Ref. [3], but with one important caveat: all examples supported only a single type of operation on encrypted data, e.g., addition or multiplication, but not both. This was a fundamental issue with the encryption schemes that were known in 1978; they all had a simple algebraic structure based on modular arithmetic, a part of which was "consumed" by the encryption scheme itself. While modular arithmetic does support additions and multiplications, the encryption layer always seemed to restrict one of these operations. This was unfortunate because an encryption scheme supporting two operations as privacy homomorphisms would enable vastly more powerful applications, as it would allow any arithmetic or Boolean circuit to be evaluated directly on encrypted data.

1.2 Homomorphic encryption

It took more than 30 years for the next big step, over which time the term *homomorphic encryption* was coined to mean encryption schemes supporting one or more privacy homomorphisms, and *fully homomorphic encryption* schemes supporting at least two privacy homomorphisms. In 2009 Craig Gentry described the first fully homomorphic encryption scheme [4] in his famous PhD dissertation; Gentry's scheme allowed both additions and multiplications—thus arbitrary circuits—to be evaluated on encrypted data. Gentry's scheme was immediately recognized as a breakthrough result in cryptography and created a massive wave of follow-up work.

However, Gentry's scheme had an extreme computational cost and was considered to be mainly of theoretical interest. In 2010–12 several new substantially simpler schemes were invented, the most famous of which are the Brakerski–Gentry–Vaikuntanathan scheme (BGV) [5] and the Brakerski–Fan–Vercauteren scheme (BFV) [6]. Other important realizations also emerged: these schemes can be perfectly functional and much more efficient in a so-called *somewhat homomorphic encryption* mode, where the schemes are parameterized to support only circuits of a predetermined maximal size. Gentry's scheme was able to support arbitrary computations by combining somewhat homomorphic encryption with a *bootstrapping* operation, where the computational capabilities of the somewhat homomorphic encryption scheme are reset at regular intervals. The problem with bootstrapping is that it is typically extremely costly compared to the normal arithmetic or Boolean operations; this is why it is often much more efficient to run schemes only in the somewhat homomorphic mode.

From 2013 onwards, several new types of homomorphic encryption schemes were invented, many of which have improved bootstrapping performance [7,8], support computing on huge encrypted integers [9], or support computing on vectors of encrypted complex numbers [10].

1.3 Note about terminology

Today, the term *fully homomorphic encryption* has an unfortunate and confusing dual meaning. It can refer to encryption schemes that support two privacy homomorphisms as opposed to one, but is also used to differentiate between bootstrappable schemes and somewhat homomorphic schemes. Encryption schemes supporting only one privacy homomorphism are commonly referred to as *partially homomorphic encryption* schemes; these tend to always allow an arbitrary amount of computation and require no bootstrapping. There are also numerous other technical descriptors used to differentiate between different types of homomorphic encryption schemes, but these are irrelevant for the present work. For more information we refer the reader to Ref. [11].

In this chapter we will simplify this terminology considerably and use the term *homomorphic encryption* to mean encryption supporting at least two privacy homomorphisms to allow arbitrary arithmetic or Boolean circuits to be evaluated on encrypted data. We also implicitly assume that the sizes of all encrypted computations are known and bounded beforehand, so that somewhat homomorphic encryption with an appropriate parameterization can always be used instead of a potentially costlier fully homomorphic encryption scheme with bootstrapping.[1]

1.4 Implementations

The first performant and still active implementation of fully homomorphic encryption was created by Shai Halevi and Victor Shoup. Their HElib library [12] achieved groundbreaking performance results in its implementation of the BGV encryption scheme. The library also featured fast algorithms for homomorphic operations, a working implementation of the complicated bootstrapping operation, and many other novel innovations supporting new types of uses of homomorphic encryption. HElib is still, at the time of writing this, the only library to implement bootstrapping for the BGV scheme.

Today, there are also many implementations with various pros and cons, depending on what the user is looking for. For example, Microsoft SEAL [13] implements the BFV and the Cheon-Kim-Kim-Song (CKKS) [10] encryption schemes,

---

[1] This is largely a matter of practicality: bootstrapping is often slow, requires parameterizations that restrict other useful properties of the encryption scheme, and is available only in few implementations.

PALISADE provides a powerful framework for implementing cryptographic primitives based on lattices,[2] and TFHE [8] allows the evaluation of arbitrary Boolean circuits on encrypted data through a fast bootstrapping operation.

1.5 Standardization

The most interesting applications of homomorphic encryption deal with extremely private—e.g., biomedical, financial, or location—data. Both from a legal and from a practical point of view the users would want assurance that the implemented encryption scheme truly is secure, that the implementation is correct and secure, and that engineers responsible for building applications are able to do it correctly and securely. For these reasons, in 2017 several key players in the field decided to initiate an open standardization consortium: HomomorphicEncryption.org.[3] This effort remains open and has grown to involve participants from several large international corporations, start-ups, governmental organizations, and academia. The most important product of this effort so far is the *Homomorphic Encryption Security Standard* [14], that describes parameterizations for homomorphic encryption that are considered to be secure against all known attacks; many libraries and research papers already implement and cite this standard.

1.6 Applications

Over the course of its history, homomorphic encryption has been applied to numerous theoretical constructions in cryptography, as well as to practical protocols. Notable early works in the applied direction include [15,16]. In 2016 the *CryptoNets* paper [17] spawned an entire line of work in evaluating deep neural networks on encrypted data, thus enabling what one might call *private predictive services*. For example, it can be possible to evaluate medical predictive models on encrypted patient data, circumventing issues related to medical data privacy, while still providing a valuable predictive service.

Since 2015 the *iDASH competition*[4] has included homomorphic encryption challenges related to genomic privacy: in 2015 the two challenges involved privacy-preserving GWAS and evaluating a distance metric on encrypted databases of SNPs, in 2016 the challenge was to enable privacy-preserving queries on encrypted outsourced genomic databases, in 2017 it was to train a predictive logistic regression model on encrypted genomic data, and in 2018 it was to learn a larger GWAS model on encrypted genomic data. Overall, the iDASH competition tasks have resulted directly and indirectly in dozens of research papers and motivated significant new work in the field.

---

[2] Most homomorphic encryption schemes are based on lattices.

[3] Accessed on February 10, 2019.

[4] http://humangenomeprivacy.org (accessed on February 10, 2019).

2. Homomorphic encryption

In this section our goal is to give the reader an understanding of how homomorphic encryption can work. For most people, hearing about an encryption scheme that allows computation to be done on encrypted data sounds impossible and should raise red flags about information leakage. However, no such leakage takes place; instead, due to the encryption scheme being heavily algebraic in nature, the computations can be thought of as passing *through* the encryption layer to the underlying plaintext data, yielding an encrypted result. Thus, it is incorrect to think that the computations would somehow extract information from the ciphertexts.

However, since most readers are probably not experts in cryptography, we start with a very high-level nonmathematical explanation of some basic concepts in the context of traditional (not homomorphic) encryption.

2.1 What is encryption?

An *encryption scheme* typically consists of at least three algorithms: key generation, encryption, and decryption. Encryption converts *plaintexts* into *ciphertexts*, and decryption has the inverse effect.

2.2 Partially homomorphic encryption

We mentioned above that the algebraic properties of homomorphic encryption schemes are what give them their incredible properties. To explain what we mean by this, we will briefly discuss how the RSA encryption scheme [18] achieves its partially homomorphic properties.

The RSA scheme works as follows. The first party, *Alice*, chooses a public modulus $N = pq$, where p and q are secret large prime numbers. Typically N is at least 2048 bits long, and finding large prime factors for such a large number is considered to be infeasible. Alice chooses also a pair of numbers e (*public exponent*) and d (*secret exponent*) such that $ed \cong 1 (\text{mod } N)$. Computing d given e is infeasible unless N can be factored. Thus, as long as p and q remain secret, revealing e does not reveal d. Alice gives the second party, *Bob*, the number N and the public exponent e.

To encrypt a message $m(\text{mod } N)$, Bob can compute $m^e(\text{mod } N)$ and send it to Alice. Alice can decrypt it by computing $(m^e)^d = m^{ed} = m(\text{mod } N)$, which requires her to know the secret exponent d. Note that only someone who knows N and d can decrypt messages, but anyone with N and e can encrypt; thus, we call (N, d) the *secret key* and (N, e) the *public key*. Encryption schemes with such a separation of keys are referred to as *public key encryption*, whereas *symmetric key encryption* refers to encryption schemes where the same key is used both for encryption and decryption. Public key encryption schemes, like RSA, are typically mathematical and algebraic in nature, while symmetric key encryption schemes are commonly based on complex ways of scrambling bits of the data according to a permutation derived from the secret key.

Now, why is RSA partially homomorphic? Suppose Alice encrypts two messages, m_1 and m_2, to obtain ciphertexts $c_1 = m_1^e \pmod N$ and $c_2 = m_2^e \pmod N$, and hands c_1 and c_2 to Bob. Since Bob does not have the secret key, he cannot recover m_1 or m_2. However, Bob can still compute the product

$$c_1 \cdot c_2 = m_1^e \cdot m_2^e = (m_1 m_2)^e \pmod N$$

without knowing or finding any secret information. If Bob gives this result to Alice, and Alice decrypts it, she will obtain the product $m_1 m_2 \pmod N$. Thus, Alice has securely outsourced the computation of the product to Bob, but Bob has learned neither the inputs, nor the output, of the computation.

Why did this work? Because the *plaintext space* (integers modulo N) is algebraic in nature, and the encryption scheme preserves some of this structure. Unfortunately, there is no way to extend this to also support addition:

$$c_1 + c_2 = m_1^e + m_2^e \neq (m_1 + m_2)^e (\mathrm{mod} N) .$$

It is now also easier to understand why creating fully homomorphic encryption is so difficult: encryption must be done in a way that is secure, yet does not ruin any of the algebraic structure of the messages.

2.3 Mathematical background

Most homomorphic encryption schemes are heavily based on the algebraic properties of certain *polynomial quotient rings*. In this section we will briefly describe the necessary mathematical background to understand how some homomorphic encryption schemes work. For more in-depth explanation of the concepts introduced here we refer the reader to any undergraduate textbook on abstract algebra.

Given a set S, $a \in S$ denotes that S contains an element a; conversely, $a \notin S$ denotes that S that S does not contain a.

We denote the set of integers by $\mathbb{Z}$ and the set of all polynomials with integer coefficients by $\mathbb{Z}[x]$. For example, $-3x^2 + 2x + 1$, $x^{1000} - 1$, 5 are all polynomials in $\mathbb{Z}[x]$. The degree of a polynomial is the highest power of x appearing in the expression; for example, the degrees of the above polynomials are 2, 1000, and 0, respectively. Polynomials in $\mathbb{Z}[x]$ can be added (coefficient-wise), and multiplied. In abstract algebra terminology, $\mathbb{Z}[x]$ forms a *ring*. The set of rational numbers is denoted by $\mathbb{Q}$ and we can also define the ring of rational coefficient polynomials: $\mathbb{Q}[x]$. Note that $\mathbb{Z}$ is a subset of $\mathbb{Q}$, and likewise $\mathbb{Z}[x]$ is a subset of $\mathbb{Q}[x]$. In fact, these are not just subsets but also *subrings*, although this distinction is not important to us. For a polynomial $p(x)$[5] we denote the i-th coefficient by $p[i]$. For example, if $p(x) = 3x^2 + 2x + 1$, $p[0] = 1$, $p[1] = 2$, and $p[2] = 3$.

Given a positive integer k, we denote the set of integers modulo k by $\mathbb{Z}_k$. Note that also $\mathbb{Z}_k$ forms a (finite) ring because it is possible to both add and multiply integers

[5] For reasons of notational simplicity we often omit the variable and denote $p(x)$ by p.

modulo k. We say that two integers a and b are equal modulo k if they differ by a multiple of k; in this sense, $\mathbb{Z}_k$ is the set of integers up to equality modulo k. For example, if the integers a and b differ by a multiple of k, then a and b denote the same element of $\mathbb{Z}_k$. For example, let $k = 5$; then every integer is equivalent to either 0, 1, 2, 3, or 4, modulo 5, and these five numbers form *a complete set of representatives* for $\mathbb{Z}_5$ with addition and multiplication working as in Table 5.1. On the other hand, any k consecutive integers form a complete set of representatives for $\mathbb{Z}_k$; for example, the integers $-2, -1, 0, 1, 2$ form another complete set of representatives for $\mathbb{Z}_5$. Given an element $a \in \mathbb{Z}_k$, we denote by $[a]_k \in \mathbb{Z}$ the integer equivalent to a modulo k in the symmetric interval

$$-\lfloor k/2 \rfloor, \ldots, -1, 0, 1, \ldots, \lfloor (k-1)/2 \rfloor,$$

where $\lfloor . \cdot \rfloor$ denotes rounding down.

We denote the set of all polynomials with coefficients in $\mathbb{Z}_k$ by $\mathbb{Z}_k[x]$. Such polynomials can again be added and multiplied as polynomials in $\mathbb{Z}[x]$, except that arithmetic for the coefficients follows that of $\mathbb{Z}_k$.

Now, let n be a power of 2.[6] We denote $R = \mathbb{Z}[x]/(x^n + 1)$; this is a so-called *polynomial quotient ring* and denotes the set of integer–coefficient polynomials of degree up to $n - 1$. Two polynomials in R can be added coefficient-wise and multiplied by changing every x^n into a-1; these operations endow the set R with a ring structure. We denote similarly $R_k = \mathbb{Z}_k[x]/(x^n + 1)$, where now the polynomial coefficients are instead integers modulo k. For example, if $n = 8$ and $k = 5$, then

Table 5.1 Addition and multiplication tables in $\mathbb{Z}_5$.

Add	0	1	2	3	4
0	0	1	2	3	4
1	1	2	3	4	0
2	2	3	4	0	1
3	3	4	0	1	2
4	4	0	1	2	3
Multiply	0	1	2	3	4
0	0	0	0	0	0
1	0	1	2	3	4
2	0	2	4	1	3
3	0	3	1	4	2
4	0	4	3	2	1

[6] This restriction is common in homomorphic encryption today and is motivated by deeper algebraic reasons that are not feasible to describe here.

$$(2x^7 - x^2 + 1) + (-2x^5 - 2x^2 + 1) = 2x^7 - 2x^5 + 2x^2 + 2$$

and

$$\begin{aligned}(2x^7 - x^2 + 1)\cdot(-2x^5 - 2x^2 + 1) &= x^{12} + x^9 - x^7 + 2x^4 + 2x^2 + 1 \\ &= -x^7 + x^4 + 2x^2 - x + 1\,.\end{aligned}$$

Note how we used the symmetric representatives of $\mathbb{Z}_5$ for the coefficients of the polynomials, e.g., $x^{12} = x^4 \cdot x^8 = -x^4$. Both R and R_k are standardized notations in homomorphic encryption literature, as these rings are in a central role in many homomorphic encryption. Understanding how they work is fundamental to understanding the rest of this chapter.

Let $p(x) \in \mathbb{Q}[x]$; its *infinity norm* $||p||$ is defined to be the absolute value of the largest coefficient:

$$||p|| = \max_j |p[j]|.$$

For $r(x) \in \mathbb{Z}_k[x]$, we define the infinity norm $||r||$ to mean the infinity norm computed for the polynomial in $\hat{r}(x) \in \mathbb{Z}[x]$, whose coefficients are the symmetric representatives of the coefficients of $r(x)$:

$$\begin{aligned}r(x) &= r_\ell x^\ell + \ldots + r_1 x + r_0\,, \\ \hat{r}(x) &= [r_\ell]_k x^\ell + \ldots + [r_1]_k x + [r_0]_k\,.\end{aligned}$$

In other words,

$$||r|| = ||\hat{r}|| = \max_j \left|[r_j]_k\right|\,.$$

2.4 (Ring) Learning With Errors

Just like RSA is secure because factoring very large numbers into large prime factors is hard, the security of the most successful homomorphic encryption schemes depends on the hardness of solving the so-called *Learning With Errors* problem (LWE), or its variant *Ring Learning With Errors* (RLWE). LWE and RLWE are computational problems that have been extensively studied for several years; solving them is estimated to be extremely hard in appropriately "large" settings. In this section we give an informal high-level description of these problems to motivate the definition of the homomorphic encryption scheme we present later later on.

2.4.1 Learning With Errors

The LWE problem was introduced by Oded Regev in Ref. [19] and can be thought of as solving a set of approximate linear equations over $\mathbb{Z}_q$, where q is a not too large integer.

In elementary linear algebra students learn to solve systems of linear equations like

$$\begin{cases} A_{0,n-1}s_{n-1} + \ldots + A_{0,1}s_1 + A_{0,0}s_0 & = b_0 \\ A_{1,n-1}s_{n-1} + \ldots + A_{1,1}s_1 + A_{1,0}s_0 & = b_1 \\ \vdots \qquad\qquad \vdots & \vdots \\ A_{m,n-1}s_{n-1} + \ldots + A_{m,1}s_1 + A_{m,0}s_0 & = b_m \end{cases}$$

using techniques such as *Gaussian elimination*. This is no more difficult even if the coefficient matrix **A** and the solution vector **s** consists of elements of $\mathbb{Z}_q$. However, now consider a variant of this system where each equation contains a small extra additive factor:

$$\begin{cases} A_{0,n-1}s_{n-1} + \ldots + A_{0,1}s_1 + A_{0,0}s_0 + e_0 & = b_0 (\text{mod } q) \\ A_{1,n-1}s_{n-1} + \ldots + A_{1,1}s_1 + A_{1,0}s_0 + e_1 & = b_1 (\text{mod } q) \\ \vdots \qquad\qquad \vdots & \vdots \\ A_{m-1,n-1}s_{n-1} + \ldots + A_{m-1,1}s_1 + A_{m-1,0}s_0 + e_{m-1} & = b_{m-1} (\text{mod } q) \end{cases} \quad (5.1)$$

Suddenly this system becomes incredibly difficult to solve when n is large, the coefficient matrix **A** is random, $n \approx m$, and q is not too large. The problem is that Gaussian elimination stops working: as soon as any of the equations if multiplied by a large coefficient, as is done in Gaussian elimination, the *error terms* e_i—even if originally small—become significant. This is the LWE problem: solving for **s** given the matrix **A** and the vector **b**.

2.4.2 Ring Learning With Errors

The RLWE problem introduced in Ref. [20] is a close relative of the LWE problem and results in improved performance for the homomorphic encryption schemes, which is why it is often used instead of LWE.

Here we present only one very specific version of the RLWE problem, again at a very high level. As in the LWE problem, we let n denote a positive integer, but in this case restrict it to be a power of 2. Let R_q be the ring $\mathbb{Z}_q[x]/(x^n + 1)$, as described above in Section 2.3. Now, consider a single polynomial equation[7]:

$$a(x)s(x) + e(x) = b(x) \ ,$$

where $a(x) \in R_q$ is random and $e(x)$ has small coefficients. Here the polynomial $s(x) \in R_q$ is unknown and the challenge to solve it is called the RLWE problem. This problem is again extremely difficult and closely related to LWE. Namely, if we expand out the product $a(x)s(x)$, the coefficients of the polynomial equation form a system of linear equations similar to what we saw in (5.1):

---

[7] More accurately, we should allow a system of multiple such equations sharing a single unknown polynomial $s(x)$. We use only a single equation here for simplicity of the exposition.

$$\begin{cases} -a_1 s_{n-1} - \ldots - a_{n-1}s_1 + a_0 s_0 + e_0 & = b_0 (\text{mod } q) \\ -a_2 s_{n-1} - \ldots + a_0 s_1 + a_1 s_0 + e_1 & = b_1 (\text{mod } q) \\ \quad \vdots \qquad\qquad \vdots & \quad \vdots \\ -a_{n-1}s_{n-1} + \ldots + a_{n-3}s_1 + a_{n-2}s_0 + e_{n-2} & = b_{n-2} (\text{mod } q) \\ a_0 s_{n-1} + \ldots + a_{n-2}s_1 + a_{n-1}s_0 + e_{n-1} & = b_{n-1} (\text{mod } q) \end{cases} \tag{5.2}$$

We have denoted $s_j = s[j]$, $a_j = a[j]$, $e_j = e[j]$, and $b_j = b[j]$, to make the notation less cluttered. If LWE can be solved, then certainly this problem can also be solved, as it is just a particular instance of LWE. While the converse is not exactly true, there is strong reason to believe that this problem is indeed extremely hard and in particular comparable to LWE when n is large enough and q is small enough [20].

2.4.3 Decision problems

The hard problems we described above are more precisely something called *search*-(R)LWE. In practice, the security of the encryption schemes rely on a seemingly weaker so-called *decision*-(R)LWE problem, which nevertheless turns out to be basically equally hard as the search problems; we refer the reader to Refs. [20,21] for full details.

The decision-(R)LWE problems are easy to describe: consider the systems in (1) and (2) and imagine that you only see the coefficient matrix $\mathbf{A}$ and the vector $\mathbf{b}$ for LWE, or the polynomial $a(x)$ and the polynomial $b(x)$ for RLWE, and that your task is not to solve the system but to merely distinguish whether what you were given truly comes from an LWE or RLWE problem, or whether you were simply given random data of the same size. If distinguishing between these two cases is difficult, then the data $(\mathbf{A}, \mathbf{b})$ (LWE) and $(a(x), b(x))$ (RLWE) are indistinguishable from random data of the same size. (R)LWE-based homomorphic encryption schemes use such data to mask plaintext elements; assuming the hardness of the decision-(R)LWE problems, the encryption is then indistinguishable from random and reveals no information about the underlying plaintext.

In practice, both $\mathbf{s}$ and $s(x)$ are often chosen to have small coefficients; this results in performance improvements in homomorphic encryption, and there is reason to believe that it does not reduce the hardness of the (R)LWE problems.

2.5 The Brakerski–Fan–Vercauteren scheme

In this section we describe the so-called Brakerski–Fan–Vercauteren (BFV) homomorphic encryption scheme [6]. Even though the BFV scheme is in general not computationally the fastest, it has gained a lot of popularity recently due to its relative simplicity in use.[8] Due to this simplicity, the BFV scheme is the only homomorphic encryption scheme we describe in any detail in this chapter.

[8] Unfortunately, this simplicity is not reflected in implementation: efficient implementations of the BFV scheme are very complicated.

We use the notation $R = \mathbb{Z}[x]/(x^n + 1)$ where n is a power of 2 and $R_k = \mathbb{Z}_k[x]/(x^n + 1)$. We occasionally also need polynomials with rational coefficients, and denote $R^{\mathbb{Q}} = \mathbb{Q}[x]/(x^n + 1)$. The polynomial $x^n + 1$ is called the *polynomial modulus*.

2.5.1 Plaintexts and ciphertexts

In the BFV scheme a plaintext is a polynomial in the ring R_t for some positive *plaintext modulus t*. A BFV ciphertext is a pair of polynomials in $R_q \times R_q$, where q is called the *ciphertext modulus*. Encryption and decryption convert plaintexts to ciphertexts and vice versa. It is necessary that $t < q$, but in practice t is much smaller than q.

It is important to understand what the meanings of the plaintext and ciphertext moduli are. The ciphertext modulus is related to the RLWE problem that provides security to the encryption scheme, whereas the plaintext modulus determines the plaintext space of the scheme. As we will see later, increasing the plaintext modulus will reduce the encrypted computation capabilities of the scheme, whereas increasing the ciphertext modulus will increase the encrypted computation capabilities of the scheme. Unfortunately, increasing the ciphertext modulus will also reduce the security of the scheme, which can subsequently be countered by increasing n.

2.5.2 Encryption and decryption

We describe both a *secret key* and a *public key* variant of BFV encryption. In both variants the secret key `sk` is generated by sampling a polynomial s from the set R so that its coefficients are sampled uniformly at random from the set $\{-1, 0, 1\}$, and setting

$$\mathrm{sk} = (1, s) \in R_q \times R_q \,.$$

For the public key variant of BFV, the public key `pk` is generated from `sk` by first sampling a polynomial a from uniformly at random, an error polynomial e with small coefficients (recall Section 2.4), and finally setting

$$\mathrm{pk} = \left([-(as + e)]_q, a\right) \in R_q \times R_q \,.$$

Here $[-]_q$ denotes modular reduction of polynomial coefficients to a symmetric interval modulo q, as was described in Section 2.3.

Let $\Delta = \lfloor q/t \rfloor$ and denote $r_t(q) = q - \Delta t$. The symmetric key variant of the BFV scheme encrypts a plaintext $m \in R_t$ by sampling a from R_q uniformly at random, an error polynomial e with small coefficients, and setting

$$\mathrm{ct} = \left([\Delta m - (as + e)]_q, a\right) \in R_q \times R_q \,. \tag{5.3}$$

Decryption works by computing

$$\widetilde{m} = \left[\left\lfloor \frac{t}{q} \langle ct, sk \rangle \right\rceil\right]_t \in R_t \,. \tag{5.4}$$

Since

$$\langle \text{ct}, \text{sk} \rangle = \Delta m - (as + e) + as + kq = \Delta m - e + kq \in R$$

for some polynomial k, we get

$$\frac{t}{q} \langle \text{ct}, \text{sk} \rangle = m - \frac{r_t(q)m + te}{q} + kt \in R^{\mathbb{Q}} .$$

As long as q is much larger than the coefficients of the polynomial $r_t(q)m + te$, rounding to the nearest integer $(\lfloor - \rceil)$ gets rid of the fractional term, leaving only $m + kt$. Thus, decryption will output $\widetilde{m} = m \in R_t$.

The public key variant of BFV is only slightly more complicated. Let $\text{pk} = (p_0, p_1)$. Encryption works by sampling a "ephemeral key" u from R with coefficients uniform in $\{-1, 0, 1\}$, error polynomials e_0, e_1 with small coefficients, and setting

$$\text{ct} = \left([\Delta m + (p_0 u + e_0)]_q, [p_1 u + e_1]_q \right) \in R_q \times R_q .$$

Decryption works as in (5.4):

$$\begin{aligned} \langle ct, sk \rangle &= \Delta m + (p_0 u + e_0) + (p_1 u + e_1)s + kq \\ &= \Delta m - asu - eu + e_0 + aus + e_1 s + kq \\ &= \Delta m - eu + e_1 s + e_0 + kq \in R , \end{aligned}$$

so

$$\frac{t}{q} \langle \text{ct}, \text{sk} \rangle = m - \frac{r_t(q)m + teu - te_1 s - te_0}{q} + kt \in R^{\mathbb{Q}} .$$

Again, as long as q is much larger than the coefficients of the polynomial $r_t(q)m + teu - te_1 s - te_0$, rounding to the nearest integer gets rid of the fractional term, leaving only $m + kt$. Thus, decryption again correctly outputs $\widetilde{m} = m \in R_t$.

2.5.3 Security

We now present a simple informal argument for the security of the BFV scheme. A formal and complete security argument would require a little bit more terminology so we omit it.

In the symmetric key setting the idea is very simple: the message m in (5.3) is (after scaling by the integer Δ) masked by subtracting from it one half of an RLWE sample. The RLWE hardness assumption guarantees that the pair $(as + e, a)$ is indistinguishable from random data of the same size, hence after masking the message becomes similarly indistinguishable from random. In decryption the randomness is trivial to remove with the help of the secret key.

In the public key variant we first argue that the public key is indistinguishable from random based on the RLWE hardness assumption. Next, the first part of the public key can be used as the coefficient polynomial in a new RLWE sample $(p_o u + e_0, p_0)$, the first part of which is used to mask the message. The second part of the ciphertext is one half of the RLWE sample (p_1u+e_1, p_1) and hence indistinguishable from random. Note that when the message $m = 0$, the ciphertext actually contains two RLWE samples with the same ephemeral key u, but with different coefficient polynomials. Thus, we in fact need the RLWE problem to remain hard given two such samples generated from a single secret. This is assumed to be the case [20].

2.5.4 Noise

In this section we introduce the fundamental concept of *noise* which is to some extent common to all modern homomorphic encryption schemes. Every ciphertext in the BFV scheme carries with itself an error term (noise), that grows in homomorphic operations. In leveled fully homomorphic encryption the encryption parameters determine an upper bound for the noise, and once this bound is reached the ciphertext becomes corrupted and impossible to decrypt, even with the correct secret key. In true fully homomorphic encryption a special *bootstrapping* process must be carried out to refresh the noise before more computations can be performed [4], but this is typically very costly. Noise typically grows very little in homomorphic additions, but grows a lot in homomorphic multiplications. Thus, typically the difficulty of performing a computation on homomorphically encrypted data is measured in terms of the multiplicative depth of the arithmetic circuit to be evaluated. We follow Ref [9] in our presentation of noise fundamentals.

Let ct $= (c_0, c_1)$ be a BFV ciphertext encrypting the message $m \in R_t$. Its *invariant noise polynomial* $v \in R^{\mathbb{Q}}$ is the polynomial with the smallest infinity norm such that

$$\frac{t}{q}\langle ct, sk\rangle = m + v + At \in R^{\mathbb{Q}}$$

for some polynomial $A \in R$. Intuitively, the invariant noise polynomial captures the term that must be rounded away in the decryption algorithm (5.4) to recover the correct message.

The above is made concrete by the following statement: a BFV ciphertext `ct` encrypting a message m decrypts correctly as long as the invariant noise polynomial v satisfies $||v|| < 1/2$. To prove this, compute

$$\tilde{m} = \left[\left\lfloor \frac{t}{q}\langle ct, sk\rangle \right\rceil\right]_t = \left[\left\lfloor \frac{t}{q}\langle ct, sk\rangle \right\rceil + kt\right]_t = \left[\left\lfloor \frac{t}{q}\langle ct, sk\rangle \right\rceil\right]_t .$$

By the definition of v,

$$\tilde{m} = \left[\lfloor m + v + At \rceil\right]_t = m + \lfloor v \rceil (\mathrm{mod}\, t) .$$

Hence, decryption is successful as long as v is removed by the rounding, i.e., when $v < 1/2$. Since only the infinity norm of the invariant noise polynomial matters in the end, we call v the *invariant noise* of `ct`. Finally, it is convenient to define a *noise budget* as $-\log_2(2||v||)$; when the noise budget reaches zero the ciphertext is no longer decryptable.

In practice we call the invariant noise polynomial simply the *noise polynomial*, and the invariant noise simply the *noise*.

2.5.5 Addition

Addition of ciphertexts in the BFV scheme is very easy. Let $ct_1 = (c_0, c_1)$, $ct_2 = (d_0, d_1)$ denote BFV ciphertexts encrypting messages m_1 and m_2 with noise polynomials v_1 and v_2, respectively. Let

$$ct_{\text{add}} = ct_1 + ct_2 = (c_0 + d_0, c_1 + d_1) \in R_q \times R_q \,;$$

then ct_{add} decrypts to $m_1 + m_2 \in R_t$ as long as $\|v_1 + v_2\| < 1/2$. To see this, note that

$$\begin{aligned}\frac{t}{q}\langle ct_{\text{add}}, sk\rangle &= \frac{t}{q}\langle ct_1, sk\rangle + \frac{t}{q}\langle ct_2, sk\rangle \\ &= m_1 + v_1 + A_1 t + m_2 + v_2 + A_2 t \\ &= (m_1 + m_2) + (v_1 + v_2) + At \in R^{\mathbb{Q}}\end{aligned}$$

for some polynomials A_1 and A_2, and $A = A_1 + A_2$. Thus, ct_{add} decrypts to $m_1 + m_2$ as long as $\|v_1 + v_2\| < 1/2$.

2.5.6 Multiplication

Multiplication is more complicated and we describe it in two parts. The first part produces a *three-component* ciphertext and the second part—often called *relinearization*—reduces the size of the ciphertext down to the usual two. Let ct_1 and ct_2 be as above, and set

$$\widetilde{ct}_{\text{mul}} = \left(\left[\left[\frac{t}{q} c_0 d_0\right]\right]_q, \left[\left[\frac{t}{q}(c_0 d_1 + c_1 d_0)\right]\right]_q, \left[\left[\frac{t}{q} c_1 d_1\right]\right]_q \right) \in R_q \times R_q \times R_q \,.$$

First we show that $\widetilde{ct}_{\text{mul}}$ decrypts to $m_1 m_2 \in R_t$ under the *extended secret key* $\widetilde{sk} = (1, s, s^2) \in R_q \times R_q \times R_q$. To see this, let

$$\begin{aligned}\frac{t}{q}\left\langle \widetilde{ct}_{\text{mul}}, \widetilde{sk}\right\rangle &= \frac{t}{q}\left\{ \left[\left[\frac{t}{q} c_0 d_0\right]\right]_q + \left[\left[\frac{t}{q}(c_0 d_1 + c_1 d_0)\right]\right]_q s + \left[\left[\frac{t}{q} c_1 d_1\right]\right]_q s^2 \right\} \\ &= \frac{t}{q}\left\{ \frac{t}{q}\left(c_0 d_0 + (c_0 d_1 + c_1 d_0)s + c_1 d_1 s^2\right) + \left(\varepsilon_0 + \varepsilon_1 s + \varepsilon_2 s^2\right) + A_0 q \right\} \in R^{\mathbb{Q}}\end{aligned}$$

where ε_i are rounding errors with $\|\varepsilon_i\| \le 1/2$, and $A_0 \in R$ is some polynomial. We denote $E(s) = \varepsilon_0 + \varepsilon_1 s + \varepsilon_2 s^2$. Now we notice that

$$c_0 d_0 + (c_0 d_1 + c_1 d_0)s + c_1 d_1 s^2 = \langle ct_1, sk\rangle \langle ct_2, sk\rangle \,,$$

so

$$\begin{aligned}\frac{t}{q}\left\langle \widetilde{ct}_{\text{mul}}, \widetilde{sk}\right\rangle &= \frac{t}{q}\langle ct_1, sk\rangle \cdot \frac{t}{q}\langle ct_2, sk\rangle + \frac{t}{q}E(s) + A_0 t \\ &= (m_1 + v_1 + A_1 t)(m_2 + v_2 + A_2 t) + \frac{t}{q}E(s) + A_0 t \\ &= m_1 m_2 + \widetilde{v}_{\text{mul}} + At \,,\end{aligned}$$

where

$$\widetilde{v}_{\text{mul}} = m_1 v_2 + m_2 v_1 + v_1 v_2 + \frac{t}{q} E(s)$$

and $A \in R$ is a polynomial. Therefore,

$$\left[\left\lfloor \frac{t}{q} \left\langle \widetilde{ct}_{\text{mul}}, \widetilde{sk} \right\rangle \right\rceil\right]_t = m_1 m_2 \in R_t \, ,$$

i.e., $\widetilde{ct}_{\text{mul}}$ decrypts to $m_1 m_2$ under the extended secret key $\widetilde{sk}$, as long as $\|\widetilde{v}_{\text{mul}}\| < 1/2$. Since $m_1 v_2$ and $m_2 v_1$ are typically the dominant terms in $\widetilde{sk}$ in the sense of $\| - \|$, it is instructive to think that noise grows by a multiplicative factor proportional to $\max\{\|m_1\|, \|m_2\|\} \leq t$. Stated in terms of noise budget, this operation has consumed $\log_2 t$ bits of it.

2.5.7 Relinearization

Next we give a conceptual explanation how $\widetilde{ct}_{\text{mul}} = (c'_0, c'_1, c'_2)$ can be reduced back to a normal ciphertext of size two that can be decrypted under the normal secret key sk. The goal is clear: create a new two-component ciphertext $ct_{\text{mul}} = (c_0^{\text{relin}}, c_1^{\text{relin}})$ such that $\langle ct_{\text{mul}}, sk \rangle = \left\langle \widetilde{ct}_{\text{mul}}, \widetilde{sk} \right\rangle$. To do this, along with the secret key and the public key, a so-called *relinearization key* must be generated. As a first attempt, suppose key generation additionally samples a polynomial a from R_q and an error polynomial e with small coefficients, and outputs a pair

$$\text{rlk} = (r_0, r_1) = \left(s^2 - (as + e), a\right) .$$

Now, observe that

$$c'_0 + \left(s^2 - (as + e)\right)c'_2 + \left(c'_1 + ac'_2\right)s = \left(c'_0 + c'_1 s + c'_2 s^2\right) - ec'_2, \qquad (5.5)$$

which motivates setting

$$c_0^{\text{relin}} = c'_0 + r_0 c'_2 \text{ and } c_1^{\text{relin}} = c'_1 + r_1 c'_2 \, .$$

While this is conceptually correct, the problem is that the extra term $-ec'_2$ acquired in (5.5) is very large and instantly depletes all noise budget.

Luckily, there is a simple and well-known trick to fix this issue, but for the sake of simplicity we will not go into such technical detail here, and instead refer the interested reader to Ref. [6].

2.5.8 Implementation complexity

As we mentioned earlier, the BFV scheme has gained a lot of popularity due to its relative simplicity compared to other comparable schemes like the BGV scheme. Unfortunately, it turns out that BGV is massively faster than BFV in naive implementations. This issue was partly addressed by Bajard et al. in Ref. [22] and subsequently improved by Halevi et al. in Ref. [23] (see also [24]), bringing the practical performance of BFV close to that of BGV. While the complexity of using BFV remained mostly unchanged and good (compared to BGV), the complexity of implementation went up dramatically. Indeed, modern implementations of BFV are very complex and work quite differently from what was described above, although conceptually and from a security perspective they are equivalent.

2.6 Computing on encrypted integers

The above description of the BFV scheme was quite technical, so let us recapitulate some main points. The BFV scheme encrypts plaintext elements, which are polynomials in the ring R_t. In other words, they are typically large polynomials of degree up to $n - 1$, where n is a power of two, whose coefficients are integers modulo t. BFV then converts these plaintext polynomials into ciphertext elements, which are pairs of typically very large polynomials in R_q.

In this case computations, i.e., the addition and multiplication operations, on the ciphertext elements results in the corresponding operations being reflected on the underlying plaintext elements. However, the plaintext elements are nothing like integer or floating point numbers, and their arithmetic is very different from integer or floating point arithmetic. Indeed, it is very hard to think of biomedical or other applications where such bizarre arithmetic would be desired. However, there are multiple ways of *encoding* integers into plaintext polynomials so that the encrypted computations yield meaningful results after the decrypted result is *decoded* [25].

As the simplest possible example, suppose you want to encrypt the integers 3 and 5. Simply convert these into plaintext polynomials $3 \in R_t$ and $5 \in R_t$, i.e., constant polynomials. This works fine as long as $t > 5$. Now, computing additions and multiplications in R_t on these constant polynomials is equivalent to computing in $\mathbb{Z}_t$: the polynomials will never grow to have more than the constant coefficient. After decrypting, decoding is done by simply reading the constant coefficient of the result plaintext, and the result will make sense as an integer as long as t is large enough so that no reduction modulo t took place during the computations. For example, computing the product of our encrypted 3 and 5 will yield the correct result—as an integer—as long as $t > 15$. Of course, this is not unfamiliar to software engineers who always need to be aware of the sizes of their data types and make sure their computations do not overflow unintentionally. On encrypted data this becomes more challenging since you cannot just inspect the values that you are computing on. Instead, you will need to bound the inputs and the size of all intermediate values, and choose t accordingly.

The problem with this approach is that it is incredibly inefficient: we used only one single coefficient of the plaintext polynomial and all the other $n - 1$ coefficients were left unused. Encryption further expanded the plaintext into a substantially larger ciphertext element. In the end, the size of our data just got expanded by a factor of thousands, or even hundreds of thousands. Obviously, this is not feasible.

2.7 Batching

Instead of using just one of the plaintext coefficients for encoding integer data, as described above, how could we leverage all of the coefficients? Unfortunately, this is not straightforward at all. For example, consider plaintext polynomials $ax+b\in R_t$ and $cx+d\in R_t$. Adding them yields $(a+c)x+(b+d)\in R_t$; this could be useful, since we have now managed in one single operation to compute two additions, and our encoding was more efficient. Multiplication yields a less exciting result: $acx^2+(ad+bc)x+bd\in R_t$.

Despite the apparent challenge of leveraging the full plaintext polynomial, there *is* a way of encoding n integers modulo t into one single plaintext polynomial in R_t so that both addition and multiplication work on those n integers simultaneously; this is often referred to as *batching*, or *packing*. The benefits are enormous in situations where the encrypted computation is highly parallelizable, e.g., multiplying or adding together $1,000,000$ pairs of numbers. In this case we only need $2\times 1,000,000/n$ plaintexts, and subsequently only $2\times 1,000,000/n$ ciphertexts, to hold the encrypted data: this is an n-fold improvement over the naive approach described in Section 2.6 both in terms of data expansion and computational cost. Typically n is in the thousands or tens of thousands, so the improvement is very significant. For inherently sequential computations where such parallelism cannot be leveraged, one might want to consider other approaches instead due to the relatively high performance overhead of the BFV scheme. We will not describe in detail how batching exactly is achieved, but refer the reader to Refs. [25,26] for more information.

While batching allows us to encrypt and operate on enormous *vectors* of integers modulo t, it is not immediately clear if the values encoded into a single batched plaintext can be made to interact. Luckily, it turns out to be possible to rotate encrypted vectors and mask out encrypted values in specific locations. These operations are not cheap in terms of noise growth, however, and should be used as sparingly as possible. We will not discuss the technicalities here, but the reader should keep in mind that such operations are indeed feasible.

2.7.1 Micro-benchmarks

We will briefly present micro-benchmarks for Microsoft SEAL, reflecting the current performance on a single CPU thread for a state-of-the-art homomorphic encryption library. We present the performance for several different choices of n, reflecting execution in different application scenarios that require more (larger n and q) or less (smaller n and q) noise room to achieve correct results. We use Microsoft SEAL version 3.2.0 [13] and run the benchmarks that come with the library on a single thread with an Intel Core i7-8700K at 3.70 GHz. The library was compiled with clang++-7 using -O3 optimization level. We combine together the multiplication and relinearization times and report also amortized times per encrypted operation on integers modulo t, i.e., assuming batching is used to a full extent. The parameter q is chosen according to the homomorphic encryption security standard [14] to be as large as possible, while still guaranteeing at least 128 bits of security, i.e., q is as large as possible to maximize the amount of computation that the specific value for n can allow. The exact value for q is given automatically by Microsoft SEAL. The results, presented in Table 5.2, reveal how critically important batching is.

Table 5.2 Micro-benchmarks for the BFV scheme from Microsoft SEAL.

n	Add	Add amort.	Multiply	Multiply amort.
4,096	16 μs	4 ns	4 ms	1 μs
8,192	74 μs	9 ns	15 ms	2 μs
16,384	349 μs	21 ns	66 ms	4 μs
32,768	1846 μs	56 ns	307 ms	9 μs

2.8 Approximate arithmetic on encrypted numbers

In this subsection we will discuss a newer scheme by Cheon-Kim-Kim-Song [27], abbreviated CKKS. In a very short time this scheme has become hugely popular and successful for reasons that will become clear to the reader momentarily. However, to appreciate it, we will go back to the BFV scheme and discuss some further challenges it presents.

In Section 2.6 we discussed a simple way of encoding integers into plaintext elements, and in Section 2.7 we discussed how batching can be used to use the entire plaintext polynomial rather than just a small part of it. We also pointed out in Section 2.6 that the programmer will need to keep track of the growth of their values to ensure that there is no overflow modulo the plaintext modulus t that might yield unexpected results.

Unfortunately, modulo t overflow in the BFV scheme is vastly more problematic than data type overflow in normal unencrypted computations. The problem is that in encrypted arithmetic *there is no way to scale down*: a 32-bit register might be good for a lot of computations on unencrypted data, but since there is no way to divide in the BFV scheme we cannot scale down the numbers in ciphertexts, and in most concrete use-cases modulo t overflow will become an issue.

The CKKS scheme is remarkable because it solves this problem, although with some significant trade-offs. Recall from Section 2.5 the decryption formula

$$\frac{t}{q}\langle ct, sk\rangle = m + v + At \in R^{\mathbb{Q}} . \tag{5.6}$$

The reason we can recover $m \in R_t$ from this is because the coefficients of the noise polynomial v are supposed to be smaller than $1/2$; hence they get rounded to zero when rounding to nearest integer. On the other hand, the message coefficients are much larger: whole integers (modulo t). Therefore, the message m and the noise do not overlap, and decryption yields the correct result. In the CKKS scheme this condition is relaxed so that the noise overlaps the message to some extent that can be controlled. Therefore, it can only ever yield approximately correct results, although we can always choose parameters that move the message far away from the noise, improving accuracy at the expense of performance. In fact, the CKKS scheme admits a natural way to encode and encrypt vectors of real or complex numbers, where such approximate computations are in any case the best one could hope for. The CKKS requires batching in the sense that there is no meaningful way to encode a single real or complex number, except by explicitly not using the other slots.

The main advantage of the CKKS scheme is in its capability of rescaling values in the batching slots. This is a remarkable difference from BFV, where no such scaling is possibly, except through extremely complicated bootstrapping-like operations [28] that are currently not implemented in any publicly available homomorphic encryption library. At a very high level, this is possible in CKKS and not in BFV because the noise overlaps the message: in BFV the message always resides in the space of whole integers as in (5.6), whereas the noise always consists of small rational numbers; this separation is not scale-invariant, which is why the same approach will not work in BFV. In CKKS the message and the noise do not have similarly absolute sizes and only their relative magnitude matters. This is why both the noise and the message can be simultaneously scaled down, and the result will still make sense. Scaling down does not come for free in the CKKS scheme either, and results in the encryption parameters changing according to a so-called *modulus switch* procedure. This *chain* of parameters is finite, hence only a limited number of scalings can be done. There is a somewhat complicated bootstrapping operation that resets the modulus switching chain [29] and results in the chain being reset. However, it does not correct approximation error that has accumulated in the computations.

In Section 3 we will discuss applications where the BFV scheme works better, and also applications where the CKKS scheme is a naturally better choice.

2.9 Other schemes

We conclude this section by a very brief mention of a few other homomorphic encryption schemes.

2.9.1 Brakerski–Gentry–Vaikuntanathan

The Brakerski–Gentry–Vaikuntanathan (BGV) scheme [5] has similar properties as the BFV scheme and was implemented in the famous HElib library [12]. Due to the HElib implementation, the BGV scheme gained a lot of popularity in the earlier days of applied homomorphic encryption. The scheme has performance benefits over BFV, which were more pronounced before new algorithmic and implementation improvements to BFV were invented. The downside of BGV is that it is more difficult to implement and use optimally.

2.9.2 Torus FHE

The Torus FHE scheme [8] is implemented in the TFHE library [30] and has gained a lot of interest recently. This scheme is significant because it allows arbitrary Boolean circuits to be evaluated on encrypted data by providing a fast bootstrapping operation. The challenge of the Torus FHE scheme is its performance in arithmetic computations, where even low-precision integer arithmetic can quickly result in large Boolean circuits that are infeasible to evaluate efficiently. On the other hand, having an encrypted bit-representation of encrypted numbers enables computations like division, square root, step function (comparison), and others that are very hard to perform using any of the other schemes.

2.9.3 High-precision arithmetic on encrypted integers

The scheme of [9] provides extremely high-precision computation capabilities on encrypted integers and rational numbers, but is currently not implemented in any publicly available library. There are currently no widely interesting applications mandating such functionality, but in some application domains—like finance—very high-precision integer arithmetic might be necessary and expected.

3. Applications

In this section we describe multiple applications and high-level application categories and explain how these applications can appear in the biomedical domain.

3.1 Outsourced storage and computation

The original motivating application [3] of homomorphic encryption is in augmenting encrypted storage with secure outsourced computation capabilities. Today, with the rise of cloud computation, this is a significant and important class of applications, as it allows data owners to be assured that the service provider cannot sell or share their data. In a basic use-case the data owner chooses the computation that is to be executed by the cloud, but sometimes the computation is not exactly known by the data owner and can instead contain private information from other parties, e.g., the service provider itself.

There are several practical challenges in the outsourced storage and computation applications. First, the computation must be more or less known beforehand, which may or may not be the case, so that the encryption parameters can be set appropriately. In practice, it can be logistically inconvenient to change the parameters later on, as a full re-encryption may be required. Another challenge is that the ciphertexts are typically much larger than plaintexts; this ratio is called message expansion, and we will discuss it more in Section 4.1. Classical approaches for organizing and storing data, such as SQL databases, may be challenging to adapt to the encrypted domain. Namely, even if only one field in such a database should be encrypted, batching links together several values from this column, and therefore links together several records. If the data are to be grouped by the values in an unencrypted field, breaking this linkage, the encrypted field may become highly fragmented, losing much of the benefit of batching. Therefore, it may not be obvious at all how to organize the data in a way that is performant, flexible, and practical.

3.2 Private prediction

One important and interesting class of applications is in enabling machine learning prediction services that operate on encrypted data: such a service would not reveal the feature data to the service provider and would not require the model owner (possibly the service provider) to reveal the model to the data owner [17]. This idea has multiple interesting applications and has resulted in a significant and ongoing line of work.

3.3 Private learning

More recently multiple papers have attempted to enable learning on encrypted data [28,31–33]. Currently, state-of-the-art research work in this direction is limited to learning linear and logistic regression models, and even this has proved to be an extremely challenging task due to the multiplicative depth of the learning algorithm, the need for bootstrapping techniques, and the typically large size of training data. One additional challenge is in controlling the learning process when the model cannot be tested at will. Nevertheless, such an encrypted learning algorithm results in an encrypted model by possibly aggregating together encrypted data from multiple data sources, while at the same time enabling persistent storage for the data.

3.4 PSI and PIR

Private Set Intersection (PSI) refers to a class of cryptographic protocols where the objective is to compute the intersection of two private sets, held by distrusting parties. A PSI protocol uses cryptographic methods to find the intersection of the two sets and returns it to one or both of the parties, without either of the parties learning anything else about each other's set. Homomorphic encryption provides a powerful primitive for PSI in cases where the two set sizes are highly asymmetric: one party holds a small set and the other a much larger set. In this case, the small set can be encrypted and sent to the other party for the intersection computation. The result, still proportional to the smaller set size, is returned and decrypted.

To see how a PSI protocol can work, consider using the BFV scheme so that each item is a number modulo t, the plaintext modulus. Suppose the smaller set consists of only one item, $x \in \mathbb{Z}_t$. Suppose the other set contains items $y_1, y_2, \ldots, y_m \in \mathbb{Z}_t$. The first party encrypts its item x and sends $\mathrm{Enc}(x)$ to the second party, who proceeds to compute

$$(\mathrm{Enc}(x) - y_1)(\mathrm{Enc}(x) - y_2)\ldots(\mathrm{Enc}(x) - y_m)$$

where the subtractions can be performed by the second party optionally encrypting its inputs y_i. Most homomorphic encryption libraries support both ciphertext-ciphertext and ciphertext-plaintext computations. Once the result is returned to the first party and decrypted, it will have the value zero precisely if x matched at least one of the $y_1, \ldots, y_m$, and a nonzero value otherwise. Batching can be used to extend the test to cover multiple values $x_1, \ldots, x_n$, so we are not restricted to one single item on the smaller set side. Unfortunately, extending to much larger sets and long items is challenging and requires a number of sophisticated tricks to be applied. We refer the reader to Refs. [34,35] for more information.

Another interesting application is in Private Information Retrieval, or PIR. PIR refers to cryptographic protocols where a server holds a dataset, and a client hopes to retrieve a row corresponding to a specific index. The server is not supposed to learn which row the client retrieved, and the client is supposed to only learn the data in that particular row. Homomorphic encryption can be used as an efficient way to implement such functionality. We will not describe the details here, but refer the reader to Refs. [35,36].

3.5 Biomedical applications

A substantial amount of applied homomorphic encryption research in recent years has been motivated or accelerated by the iDASH competition. The challenges presented in the homomorphic encryption track have included topics such as private edit distance computation [37], storing and querying genomic data for the presence of specific mutations [38], and learning linear models on encrypted genomic data [28,31–33]. Other highly practical scenarios involve evaluating linear models on encrypted genomic data to provide privacy-preserving predictions. Such models can easily involve up to millions of SNPs and are typically very fast to evaluate.

Biomedical applications can be particularly interesting for multiple reasons. First, in the biomedical domain privacy concerns are particulary serious, so the need for strong encryption is clear. Second, unlike in many other application domains, biomedical applications can typically tolerate a significantly longer latency: it may not be a problem if a biomedical computation takes seconds, minutes, or even days in some cases. Still, challenges remain. Encrypted computations are not generic, and in some cases biomedical computations simply require operations cannot be reasonably approximated by moderate size arithmetic circuits, making them unsuitable for homomorphic encryption. Another potential issue is that the encrypted data size can be significantly larger than unencrypted data size; an expansion rate of under 10x should be considered good. Hence, storing very large amounts of homomorphically encrypted data may not make sense, and instead a better solution should be considered.

4. Future Outlook

So far we have mostly discussed what the benefits of homomorphic encryption are, but have ignored the multiple challenges and downsides that are involved in using this technology. In this section we start by describing some of these challenges, and continue by describing potential solutions, and other aspects that will be important when homomorphic encryption is widely deployed.

4.1 Complexity

Homomorphic encryption increases the computation complexity by typically several orders of magnitude. With most schemes, additions of encrypted integers are fast, but multiplications take typically at least a millisecond on a commodity CPU, and for the biggest meaningful parameterizations with more computational capability this time can go up to 100 milliseconds or more. Of course, if the computation is highly parallel, we can utilize batching (Section 2.7) to improve the performance by potentially several orders of magnitude. Nevertheless, only some computations can utilize batching efficiently, and this strongly restricts the types of computations that are a meaningful target for applying homomorphic encryption.

Another related issue is message expansion: ciphertext are always larger than plaintexts, and depending on the parameterization they can be significantly larger. In the context of the BFV scheme, for example, the message expansion is *ideally* $\log(t)/\log(q)$, using the notation of Section 2.5. However, this is assuming again that the plaintext element, i.e., a polynomial in R_t, is fully utilized by useful data. This is likely to be the case only when batching can be utilized to full extent, and instead in most use-cases there is an extra factor of message expansion coming from a nonideal encoding technique.

4.2 Usability

Even traditional encryption schemes tend to be very hard to use by normal software developers [39], and homomorphic encryption brings this challenge to a new level. The complexity of parameterizing homomorphic encryption schemes, designing computations that utilize batching in an optimal way, and using currently available library implementations securely poses a significant challenge to the scalability of this technology even if the performance issues were totally resolved. Only recently library developers have started paying significant attention to usability aspects, partly due to the ongoing open standardization effort discussed earlier in Section 1.5.

Even if the homomorphic encryption primitives were easy to use, it is not easy to build secure protocols from homomorphic encryption. There has so far not been much research in this direction, but recent works on building Private Set Intersection [34,35] and Private Information Retrieval [36] protocols from homomorphic encryption reveal some of these challenges and employ various ad hoc solutions. Currently, it does not seem realistic that a normal software developer would be able to safely build secure and correct homomorphic encryption-based solutions.

From the point of view of software developers, homomorphic encryption libraries can be hard to master due to the need to understand details of the scheme for optimal use and the unusual programming model of encrypted computation. For example, branching is not possible on encrypted values, computations are typically highly vectorized, and their running times can be very hard to predict due to the discrete nature of available parameter choices. When writing normal programs the programmer can expect that leaving out one instruction from a computation will not have a significant effect on performance. With homomorphic encryption, it can enable the use of smaller encryption parameters, which can significantly change the performance and message expansion of the encryption scheme.

4.3 Hardware

Hardware acceleration from either GPUs or FPGAs has already proved to be a viable way of improving the computation performance of homomorphic encryption

[40–43]. However, such implementations are complex and are not widely available today, except in research prototype libraries [41]. While it is hard to predict how the situation will exactly evolve even in the near future, it seems likely that in particular FPGAs will be developed in a direction that is beneficial for achieving far better results than what is reported publicly today. For GPUs, the hardware development is driven strongly by machine learning and gaming applications, neither of which has similar needs for vectorized high-precision integer (and floating-point [23]) capabilities that homomorphic encryption has. On the other hand, while FPGAs are a realistic solution in data center deployments, it seems unlikely that normal PCs would be running such devices in the near future, whereas powerful gaming GPUs are widely and easily available.

Another question is the feasibility of running homomorphic encryption on weak hardware, such as medical devices, sensors, cameras, and the like. The author is unaware of any public implementation or significant research work aiming at running any part of homomorphic encryption (encryption, decryption, or encrypted computation) on a device weaker than a modern smart phone, but this is certainly possible.

5. Conclusions

In this chapter we have introduced a relatively new powerful cryptographic technology called homomorphic encryption. We have described an example of a modern homomorphic encryption scheme—the BFV scheme—and discussed multiple other schemes at a high level. We have also discussed practical aspects related to using these schemes, such as encodings, batching, and hardware acceleration. We presented multiple applications at a high level, but noted that as of today, there are no known large-scale deployments of homomorphic encryption.

References

[1] HIPAA Journal. Healthcare data breach statistics. Retrieved from: https://www.hipaajournal.com/healthcare-data-breach-statistics. (Accessed on February 10, 2019).

[2] Liu V, Musen MA, Chou T. Data breaches of protected health information in the United States. JAMA 2015;313(14):1471–3.

[3] Rivest RL, Adleman L, Dertouzos ML. On data banks and privacy homomorphisms. Foundations of Secure Computation 1978;4(11):169–80.

[4] Gentry C. Fully homomorphic encryption using ideal lattices. STOC; 2009. p. 169–78.

[5] Brakerski Z, Gentry C, Vaikuntanathan V. (Leveled) fully homomorphic encryption without bootstrapping. ITCS; 2012. p. 309–25.

[6] Fan J, Vercauteren F. Somewhat practical fully homomorphic encryption. 2012. Cryptology ePrint Archive, Report 2012/144. Retrieved from: https://eprint.iacr.org/2012/144.

[7] Ducas L, Micciancio D. FHEW: bootstrapping homomorphic encryption in less than a second. In: EUROCRYPT, part I; 2015. p. 617–40.

[8] Chillotti I, Gama N, Georgieva M, Izabachène M. Faster fully homomorphic encryption: bootstrapping in less than 0.1 seconds. In: ASIACRYPT, part I; 2016. p. 3–33.
[9] Chen H, Laine K, Player R, Xia Y. High-precision arithmetic in homomorphic encryption. In: Cryptographers' track at the RSA conference. Springer; 2018. p. 116–36.
[10] Cheon JH, Kim A, Kim M, Song Y. Homomorphic encryption for arithmetic of approximate numbers. In: International conference on the theory and application of cryptology and information security. Springer; 2017. p. 409–37.
[11] Armknecht F, Boyd C, Carr C, Gjøsteen K, Jäschke A, Reuter CA, Strand M. A guide to fully homomorphic encryption. 2015. Cryptology ePrint Archive, Report 2015/1192. Retrieved from: https://eprint.iacr.org/2015/1192.
[12] Halevi S, Shoup V. Design and implementation of a homomorphic-encryption library. IBM Research (Manuscript) 2013;6:12–5.
[13] Microsoft SEAL (release 3.2.0).. February 2019. Retrieved from: http://sealcrypto.org. Microsoft Research, Redmond, WA.
[14] Albrecht M, Chase M, Chen H, Ding J, Goldwasser S, Gorbunov S, Halevi S, Hoffstein J, Laine K, Lauter K, Lokam S, Micciancio D, Moody D, Morrison T, Sahai A, Vaikuntanathan V. Homomorphic encryption security standard. Technical report. Toronto, Canada: HomomorphicEncryption.org; November 2018.
[15] Naehrig M, Lauter KE, Vaikuntanathan V. Can homomorphic encryption be practical? CCSW; 2011. p. 113–24.
[16] Gentry C, Halevi S, Smart NP. Homomorphic evaluation of the AES circuit. CRYPTO; 2012. p. 850–67.
[17] Gilad-Bachrach R, Dowlin N, Laine K, Lauter KE, Naehrig M, Wernsing J. CryptoNets: applying neural networks to encrypted data with high throughput and accuracy. ICML; 2016. p. 201–10.
[18] Rivest RL, Shamir A, Adleman L. A method for obtaining digital signatures and public-key cryptosystems. Communications of the ACM 1978;21(2):120–6.
[19] Regev O. On lattices, learning with errors, random linear codes, and cryptography. Journal of the ACM (JACM) 2009;56(6):34.
[20] Lyubashevsky V, Peikert C, Regev O. On ideal lattices and learning with errors over rings. In: Annual international conference on the theory and applications of cryptographic techniques. Springer; 2010. p. 1–23.
[21] Eisenträger K, Hallgren S, Lauter K. Weak instances of PLWE. In: International workshop on selected areas in cryptography. Springer; 2014. p. 183–94.
[22] Bajard J-C, Eynard J, Hasan A, Zucca V. A full RNS variant of FV like somewhat homomorphic encryption schemes. SAC; 2016.
[23] Halevi S, Polyakov Y, Shoup V. An improved RNS variant of the BFV homomorphic encryption scheme. 2018. Cryptology ePrint Archive, Report 2018/117. Retrieved from: https://eprint.iacr.org/2018/117.
[24] Badawi AAl, Polyakov Y, Aung KMM, Veeravalli B, Rohloff K. Implementation and performance evaluation of RNS variants of the BFV homomorphic encryption scheme. 2018. Cryptology ePrint Archive, Report 2018/589, https://eprint.iacr.org/2018/589.
[25] Dowlin N, Gilad-Bachrach R, Laine K, Lauter KE, Naehrig M, Wernsing J. Manual for using homomorphic encryption for bioinformatics. Proceedings of the IEEE 2017; 105(3):552–67.
[26] Smart NP, Vercauteren F. Fully homomorphic SIMD operations. Designs, Codes and Cryptography 2014;71(1):57–81.

[27] Cheon JH, Kim A, Kim M, Song Y. Homomorphic encryption for arithmetic of approximate numbers. In: ASIACRYPT, part I; 2017. p. 409–37.
[28] Chen H, Gilad-Bachrach R, Han K, Huang Z, Jalali A, Laine K, Lauter K. Logistic regression over encrypted data from fully homomorphic encryption. BMC Medical Genomics 2018;11(4):81.
[29] Cheon JH, Han K, Kim A, Kim M, Song Y. Bootstrapping for approximate homomorphic encryption. In: Annual international conference on the theory and applications of cryptographic techniques. Springer; 2018. p. 360–84.
[30] Chillotti I, Gama N, Georgieva M, Izabachène M. TFHE: Fast fully homomorphic encryption over the torus. 2018. Cryptology ePrint Archive, Report 2018/421, https://eprint.iacr.org/2018/421.
[31] Kim A, Song Y, Kim M, Lee K, Cheon JH. Logistic regression model training based on the approximate homomorphic encryption. BMC Medical Genomics 2018;11(4):83.
[32] Kim M, Song Y, Wang S, Xia Y, Jiang X. Secure logistic regression based on homomorphic encryption: design and evaluation. JMIR Medical Informatics 2018;6(2).
[33] Crawford JLH, Gentry C, Halevi S, Platt D, Shoup V. Doing real work with FHE: the case of logistic regression. In: Proceedings of the 6th workshop on encrypted computing & applied homomorphic cryptography. ACM; 2018. p. 1–12.
[34] Chen H, Laine K, Rindal P. Fast private set intersection from homomorphic encryption. In: Proceedings of the 2017 ACM SIGSAC conference on computer and communications security. ACM; 2017. p. 1243–55.
[35] Chen H, Huang Z, Kim L, Rindal P. Labeled psi from fully homomorphic encryption with malicious security. In: Proceedings of the 2018 ACM SIGSAC conference on computer and communications security. ACM; 2018. p. 1223–37.
[36] Angel S, Chen H, Laine K, Setty S. Pir with compressed queries and amortized query processing. In: 2018 IEEE symposium on security and privacy (SP). IEEE; 2018. p. 962–79.
[37] Cheon JH, Kim M, Lauter K. Homomorphic computation of edit distance. In: International conference on financial cryptography and data security. Springer; 2015. p. 194–212.
[38] Çetin GS, Chen H, Laine K, Lauter K, Rindal P, Xia Y. Private queries on encrypted genomic data. BMC Medical Genomics 2017;10(2):45.
[39] Green M, Smith M. Developers are not the enemy!: the need for useable security apis. IEEE Security & Privacy 2016;14(5):40–6.
[40] Dai W, Doröz Y, Sunar B. Accelerating NTRU based homomorphic encryption using GPUs. In: 2014 IEEE high performance extreme computing conference (HPEC). IEEE; 2014. p. 1–6.
[41] Dai W, Sunar B. CUHE: a homomorphic encryption accelerator library. In: International conference on cryptography and information security in the Balkans. Springer; 2015. p. 169–86.
[42] Khedr A, Gulak G, Vaikuntanathan V. Shield: scalable homomorphic implementation of encrypted data-classifiers. IEEE Transactions on Computers 2015;65(9):2848–58.
[43] Riazi MS, Laine K, Pelton B, Dai W. HEAX: High-performance architecture for computation on homomorphically encrypted data in the cloud. 2019. arXiv preprint arXiv:1909.09731.

CHAPTER

Secure multi-party computation

6

Yan Huang, PhD

Assistant Professor, Computer Science Indiana University, Bloomington, IN, United States

1. A brief overview

Human genomes (and other medical records in general) are typically regarded to be highly sensitive personal information. On the other hand, human genomes play an increasingly important role in many trades of our society including healthcare, precision medicine, and scientific studies in general. The use of personal genomes in these trades can always be abstracted as running certain computations that take the genomes as inputs. More often than not, these sensitive genomic inputs would be contributed by different sources such as individuals, hospitals, or entrusted data authorities. To put it more concretely, consider three kinds of example scenarios:

- Two or more hospitals (perhaps from very different countries/regions) would like to conduct a joint computational mining of their patient genomes to derive a more accurate model for diagnosing certain diseases.
- To improve the outcome of medical treatments, a doctor would like to query a database of patient-records for some candidates who are genetically similar to his/her customer.
- A couple of individuals may be interested in predicting the healthiness of their would-be offsprings or the genetic compatibility of their potential marriage (e.g., through leveraging the research results of behavioral genetics) before (or at the very early stage of) the dating process.

As this computation involves sensitive data from multiple sources, a conventional solution would require the actual computation to occur at a certain mutually trusted place where all sensitive inputs are gathered. This raises several outstanding security concerns: (1) The sensitive data will leave the data owner, and future access to the dataset may be out of control. (2) The place where the joint computation occurs will easily become the single point of failure, for example, the software/hardware platform could be hacked, and some insiders of the trusted entity can be corrupted. (3) It can be very expensive to employ a mutually trusted party. In some circumstances, due to the high stake in the sensitive data, it can even be impossible to establish a trusted party. Therefore it is imminent to ask, *Without a mutually trusted party, is it still possible to accomplish the joint computation while retaining full control of every party's private data?*

Responsible Genomic Data Sharing. https://doi.org/10.1016/B978-0-12-816197-5.00006-1

A whole area of modern cryptography, generally known as *secure multiparty computation* (or MPC), has been invented to answer this question. Using cryptographic MPC techniques, all secret-holding participants no longer need to use a mutually trusted third party but can rely purely on well-known mathematical hardness assumptions such as AES is a secure block cipher and the decisional Diffie–Hellman problem is hard. In return, MPC protocols are able to offer the same security guarantee as if an *invincible* trusted party was employed to conduct the computation.

2. Defining security

What security guarantee is offered by cryptographic MPC protocols? To answer this question, one needs to consider two major aspects:

- **The capability of an attacker.** Several (orthogonal) factors can affect the adversary's power. First, an adversary can be *passive*, *active*, or *covert*. A *passive adversary* has to always obey the protocol specifications but can passively observe the transcript and run side-computation over what's passively observable to break the protocol security. An *active adversary*, on the other hand, can deviate from the protocol in arbitrary ways to break security. The passive adversary model is very weak as it is applicable only when additional nontechnical means can ensure the adversaries' behavior always conforms to the protocol specification (e.g., if the protocol implementation runs inside a protected environment inaccessible to the adversary). The active adversary model, on the other hand, is well suited to capture attackers in realistic scenarios. As a common practice, protocols only secure in the presence of passive adversaries are first developed, then to derive a version of the protocol that is secure against active adversaries, extra cryptographic techniques will be devised to enforce honest behavior (in privacy-preserving ways) when executing the protocols. Lying somewhat between the two extremes, there are also *covert* adversaries who are willing to deviate from the protocol specification only when the chance of cheating behaviors being caught is lower than some predefined threshold, as it is assumed that such attackers cannot afford the (expected) penalty of being caught cheating.

Second, threat models can also be divided into groups based on which subset of players an adversary can corrupt. Take three-party secure computation as an example, protocols that are secure in the presence of only corrupting a single party can be very different from those that can withstand attacks corrupting two parties. In the general n-party computation cases, notable categories of MPC scenarios include settings (1) where any number of parties are corrupted, (2) when a minority of the parties are corrupted, and (3) when less than one-third of the parties are corrupted, by an active adversary. Note that an attacker corrupting k parties is equivalent to that k parties collude to break the security.

- **The criteria of successful attacks.** A formal threat model also has to clarify what exactly will be considered a "successful attack." This is particularly challenging for interactive protocols involving many participants. The resolution is to first define an ideal model execution where a trusted third party exists to guarantee security and then define the security of any real model execution (of peer-to-peer cryptographic protocols) by comparing the *effects* of the two execution models. Intuitively, we consider it a successful attack to real model protocols whenever the attack allows the adversary to learn anything about other player's secret that was not possible in a corresponding ideal model execution.

Motivated by various deployment scenarios, some other security and utility constraints may also come into play. For example, whether the adversary is allowed to decide whom to corrupt based on its interaction with its peers? Protocols secure against such adversaries are said to be *adaptively* secure, whereas those ignoring such attacks are only *statically* secure. Further, in some settings, we may prefer multiple participating parties to obtain their outputs from the joint computation. Then there comes a *fairness* issue, that is, whether all the parties can receive their results at the same time or not. Protocols able to ensure all parties to receive results simultaneously are said to be *fair*. Finally, some protocols are expected to run with minimal requirements on peer-to-peer interactions, which motivated research works on *noninteractive secure computation* protocols, and secure computation *without simultaneous onlineness*.

Nongoals of MPC It is important to clarify that multiparty computation cannot solve all security concerns involving sensitive human-genomic and medical data. Some security issues are certainly beyond the goals of MPC research. For instance, all secure MPC protocols guarantee nothing about the leakage through the desired outcomes of the (publicly agreed) computation. Therefore in the two-party setting, it does not make sense to engage a secure protocol to compute an edit-distance-based alignment between two secrete genomes owned by the two participating parties, as an adversary can easily recover the other party's secret genome from its own input and the resulting alignment. As a rule of thumb, the parties should only agree on computing functions whose result leak as little information as possible, for example, only outputting whether the edit-distance is within a threshold or not.

Another major concern of hospitals and research institutions that are delegated with a tremendous amount of sensitive patient records is whether (and how) is it possible to leverage their rich datasets to facilitate biomedical research but without compromising an individual's privacy. We note that multiparty computation protocols alone are not able to address this privacy issue. This first squarely in the goal of differential privacy research, while stitching MPC with the state-of-the-art differential privacy techniques may be nontrivial.

Thirdly, even if actively secure MPC protocols are employed such that one does not need to worry about *any* attacks from its peers (for example, due to incorrect or compromised software implementations), the security of the whole system still relies on the correctness of the software agent running on its own side. In this sense,

MPC techniques only help to reduce the trust-base of the whole system but cannot eliminate the trust on software implementations. Unfortunately, it is extremely difficult to translate such highly sophisticated interactive cryptographic protocols from their descriptions in academic papers to their practical software implementations. Without a faithful implementation, an MPC system can often be shown vulnerable to various attacks, or no longer known to be provably secure at the best.

The ideal model versus real model paradigm As an illustration of the basic idea of the ideal/real model paradigm, we present the security definition of MPC protocols in the two-party setting, ignoring issues such as fairness and adaptive attacks.

Consider party P_1 (with input x) and P_2 (with input y) want to jointly compute $f(x, y)$ where f is a function they agreed on. The execution in the idea model in which a mutually trusted party T exists works as follows: (1) P_1 and P_2 send x and y, respectively, to the trusted party T over two private and authenticated channels. (2) Upon receiving x, y, T computes $f(x, y)$ and sends it to both parties. Note that as we do not consider fairness issues, we simply assume the adversary-controlled party cannot abort prematurely.

Definition 2.1. For any efficient adversary A' corrupting either P_1 or P_2, define $\mathbf{IDEAL}_{f,A'}(x, y)$ be the outputs of P_1 and P_2 after executing the ideal model protocol above with input x, y being the inputs of P_1 and P_2, resp. Note that

$\mathbf{IDEAL}_{f,A'}(x_1, x_2)$ can be a distribution over the randomness of T and A'.

Similarly, we can generally define an execution in the real model as follows. Given a public distributed algorithm Π_f, P_1 and P_2 run Π_f with their respective secret inputs x, y. No premature abort is allowed to occur on the adversarial party's side.

Definition 2.2. For any efficient adversary A corrupting either P_1 or P_2, define

$\mathbf{REAL}_{\Pi_f,A}(x, y)$ be the *joint* outputs of P_1 and P_2 after executing the real model protocol Π_f with x, y being the inputs of P_1 and P_2, resp. Note that.

$\mathbf{REAL}_{\Pi_f,A}(x, y)$ can be a distribution over the randomness of P_1, P_2, and A.

Definition 2.3. (Security). The real model protocol Π_f is said to *securely realize f* if, for all efficient A, it is possible to construct an efficient S such that for all x, y,

$\mathbf{IDEAL}_{f,A'}(x, y)$ is computationally indistinguishable from $\mathbf{REAL}_{\Pi_f,A}(x, y)$.

We remark that this paradigm models what was possibly leaked to the adversary during the execution by allowing A to output whatever it wants to. The key observation is that, if regardless of what A outputs in real model executions, there also exists a similar adversary S in the ideal model who cannot launch any attack by definition but can still produce a computationally indistinguishable output, there should not be any computationally noticeable leakage.

3. Protocols

Since 1980s, researchers have proposed many MPC protocols, with various tradeoffs in terms of their computation/bandwidth/round complexities and security guarantees. Later, we will present several representative MPC protocols along with relevant optimizations.

3.1 Oblivious transfer

One-out-of-two oblivious transfer $\left(\mathrm{OT}_1^2\right)$ itself can be considered one of the simplest MPC protocols but also a crucial building block of many generic MPC protocols. Intuitively, an OT_1^2 protocol allows a *sender*, holding strings ω_0, ω_1, to transfer to a *receiver*, holding a selection bit b, exactly one of the inputs ω_b; the receiver learns nothing about ω_{1b}, and the sender does not learn b. The oblivious transfer has been studied extensively, and several protocols are known based on a variety of hardness assumptions. Notably, there are oblivious-transfer extension techniques that can achieve a virtually unlimited number of oblivious transfers at the cost of (essentially) κ executions of OT_1^2 (where κ is a statistical security parameter) plus a marginal cost of a few symmetric-key operations per additional OT. The effect is analogical to that of hybrid encryption where symmetric ciphers are used to offset the price of expensive asymmetric ciphers. Finally, general k-out-of-n oblivious transfers can also be built upon OT_1^2s.

3.2 Multiplicative triples

A multiplicative triple, also simply called *AND triple*, consists of tuples of 3 bits (a_i, b_i, c_i) held by a party P_i such that

$$(a_1 \oplus \cdots \oplus a_n) \cdot (b_1 \oplus \cdots \oplus b_n) = c_1 \oplus \cdots \oplus c_n.$$

Protocols for generating AND triples are examples of secure multiparty computation that allows the ith participants P_i $(1 \leq i \leq n-1)$ holding secret input (a_i, b_i, c_i) and P_n holding input (a_i, b_i) to compute

$$c_n : = \bigoplus_{i=1}^{n} a_i \cdot \bigoplus_{i=1}^{n} b_i \oplus \bigoplus_{i=1}^{n-1} c_i$$

and sends it only to P_n. It is well known that an AND triple enables the parties who hold the corresponding bit shares to AND three secret bits that are additively shared among the parties. As any computation can be expressed as boolean circuits of ANDs and XOR gates, and, as we will explain soon, it is relatively easy to compute XOR gates, protocols for producing AND triple are considered one of the most important primitive protocols for generic MPC schemes.

3.3 Generic MPC in linear rounds

In general, any Turing-complete computation can be expressed as boolean circuits of ANDs and XORs. A common basic idea behind many MPC protocols is for every party to turn every bit of its secret input into n shares using some secret-share scheme and distribute the ith share to P_i; then, the n parties will coordinate to run a public algorithm to compute ANDs and XORs over those predistributed shares.

In the simplest two-party case, let $a, b \in \{0, 1\}$ be two secret bits whose (uniform random) shares a_1, a_2, b_1, b_2 (such that $a_1 \oplus a_2 = a$ and $b_1 \oplus b_2 = b$) are dealt with P_1 and P_2. Obviously, it is easy to compute $a \oplus b$ by requiring P_i to compute $c_i :=$

$a_i \oplus b_i$, where c_i can be treated as the ith share of $a \oplus b$. To compute $a \oplus b$, that is, $(a_1 \oplus a_2)(b_1 \oplus b_2) = (a_1b_1 \oplus a_2b_2 \oplus a_1b_2 \oplus a_2b_1$, we note a_1b_1 and a_2b_2 can be locally derived by P_1 and P_2, respectively. Hence, it suffices to focus on the computation of $a_1b_2 \oplus a_2b_1$. One way to achieve this is through an 1-out-of-4 oblivious transfer, for example, P_1 acts as the OT sender with inputs $c_1, a_2 \oplus c_1, a_1 \oplus c_1$, and $a_1 \oplus a_2 \oplus c_1$ where c_1 is a secret uniform bit picked by P_1 , while P_2 as the OT receiver using $b_1 \| b_2$ as the two choice bits to learn $c_2 := a_1b_2 \oplus a_2b_1 \oplus c_1$. Note that each OT needs at least one round but the input to the OT of next layer's AND gate will have to depend on the output of the OT of the previous layer. Therefore, this approach, also known as Goldwasser-Micali-Wigderson (GMW) [1], requires linear rounds in terms of the depth of the circuit, even assuming there is always sufficient space to store all the intermediate gates at the same layer of AND gates.

The GMW approach can generally work with n parties. XOR gates can be computed without interaction, similar to the two-party setting. For ANDs, every pair of P_i and P_j (where $i \neq j$) needs to run an OT protocol to compute $a_ib_j \oplus a_jb_i$ and XOR-share the result between P_i (holding $c_{i,j}$) and P_j (hold $c_{j,i}$) so that $c_{i,j} \oplus c_{j,i} = a_ib_j \oplus a_jb_i$. Then, each party P_i locally computes $c_i := a_ib_j \oplus \oplus_{j=1}^{n} c_{i,j}$ as its share of $a \cdot b$ (since $\oplus c_i = \oplus a_i \cdot \oplus b_i$).

It is also possible to generalize the idea so that the secret signals represent elements of any particular finite field $\mathbb{F}$ (instead of binary field). Therefore the circuit can accomplish any computation over signals of $\mathbb{F}$. In this case, general Shamir's secret sharing scheme can be used if $|\mathbb{F}| > n$, that is, the size of $\mathbb{F}$ is greater than the number of participating parties. An extra benefit of using Shamir's scheme is that the shares can be generated to ensure that, fixing a parameter t, any subset of t or more honest parties will be able to conduct the computation. As the shares in Shamir's secret sharing scheme are homomorphic with respect to the field addition, securely addition only requires every party summing up its corresponding local shares. To compute $a \cdot b$, however, a polynomial degree reduction step will be needed. The secure degree reduction protocol requires constant rounds assuming each party has sufficient memory so that all independent messages can be sent or received within the same round. For malicious adversaries, verifiable secret sharing can be used to identify potential protocol deviations. This approach is commonly known as the BGW [2] protocol.

3.4 Generic MPC in constant rounds

Garbled circuits allow two parties holding inputs x and y, respectively, to evaluate an arbitrary function $f(x, y)$ without leaking any information about their inputs beyond what is implied by the function output. To compute an arbitrary function f using the garbled circuit, the basic idea is to let one party (called the *garbler*) prepare an "encrypted" version of the circuit computing f; the second party (called the *evaluator*) then obliviously evaluates the encrypted circuit without learning any intermediate values.

Starting with a Boolean circuit for f (agreed upon by both parties in advance), the garbler associates two random cryptographic keys L_i^0, L_i^1 (also known as *wire-labels*) for the ith wire in the circuit (L_i^0 encodes a 0-bit and L_i^1 encodes a 1-bit). Then, for each binary gate g of the circuit with input wires i, j and output wire k, the garbler computes ciphertexts $\mathrm{Enc}_{L_i^{b_i}, L_j^{b_j}}\left(L_k^{g(b_i,b_j)}\right)$ for all possible values of $b_i, b_j \in \{0,1\}$. The resulting four ciphertexts, in random order, constitute a garbled gate for g. The collection of all garbled gates forms the garbled circuit that is sent to the evaluator. In addition, the garbler reveals the mappings from output-wire keys to bits.

To start circuit evaluation, the evaluator must also obtain the appropriate keys (that is, the keys corresponding to each party's actual input) for the input wires. The generator can simply send $L_1^{x_1}, \ldots, L_n^{x_n}$, the keys that correspond to its own input where each $L_1^{x_i}$ corresponds to the generator's ith input bit. The parties use the oblivious transfer to enable the evaluator to obliviously obtain the input-wire keys corresponding to its own inputs. Given keys L_i, L_j associated with both input wires i, j of some garbled gate, the evaluator can compute a key for the output wire of that gate by decrypting the appropriate ciphertext. Thus given one key for each input wire of the circuit, the evaluator can compute a key for each output wire of the circuit. With the mappings from output-wire keys to bits provided by the garbler, the evaluator can learn the actual output of f. If desired, the evaluator can then send this output back to the circuit generator. However, note that sending the output back to the generator is a privacy risk unless the adversary is semihonest.

The point-and-permute technique The *point-and-permute* technique proposed by Pinkas et al. [3] enables the evaluator to always compute a single decryption per gate. The idea is to use L_i^0 to represent a random bit $\lambda_i \in \{0,1\}$ on the ith wire so that bit $b_i \oplus \lambda_i$ can be revealed to the evaluator to index the garbled entry for decryption. More specifically, let $\lambda_i, \lambda_j, \lambda_k$ be the *permutation* bits of the two input wires and the output wire of a gate. Then L_i^0 and L_j^0 should be used to encrypt the output key $L_k^{g(\lambda_i,\lambda_j,)\oplus\lambda_k}$. Thus a garbled table for $g =$ AND can be expressed as

$b_i \oplus \lambda_i$	$b_j \oplus \lambda_j$	$b_k \oplus \lambda_k$	Point and permute
0	0	$z_{00} = \lambda_i\lambda_j \oplus \lambda_K$	$\mathrm{Enc}_{L_i^0, L_j^0}(L_k^{z_{00}}, z_{00})$
0	1	$z_{01} = \lambda_i\bar{\lambda}_j \oplus \lambda_K$	$\mathrm{Enc}_{L_i^0, L_j^1}(L_k^{z_{01}}, z_{01})$
1	0	$z_{10} = \bar{\lambda}_i\lambda_j \oplus \lambda_K$	$\mathrm{Enc}_{L_i^1, L_j^0}(L_k^{z_{10}}, z_{10})$
1	1	$z_{11} = \bar{\lambda}_i\bar{\lambda}_j \oplus \lambda_K$	$\mathrm{Enc}_{L_i^1, L_j^1}(L_k^{z_{11}}, z_{11})$

As the evaluator does not know $\lambda_i, \lambda_j, \lambda_k$, it is safe to send the garbled table without further permutation. Further, note that in the random oracle model, $\mathrm{Enc}_{x,y}(Z)$ can be realized as $\mathrm{H}(x,y) \oplus Z$ where H is modeled as a random oracle. Hence, a garbled gate can be simply implemented as

$b_i \oplus \lambda_i$	$b_j \oplus \lambda_j$	$b_k \oplus \lambda_k$	With Random Oracle H
0	0	$z_{00} = \lambda_i \lambda_j \oplus \lambda_K$	$H\left(L_i^0, L_j^0\right) \oplus \left(L_k^{z_{00}}, z_{00}\right)$
0	1	$z_{01} = \lambda_i \bar{\lambda}_j \oplus \lambda_K$	$H\left(L_i^0, L_j^1\right) \oplus \left(L_k^{z_{01}}, z_{01}\right)$
1	0	$z_{10} = \bar{\lambda}_i \lambda_j \oplus \lambda_K$	$H\left(L_i^1, L_j^0\right) \oplus \left(L_k^{z_{10}}, z_{10}\right)$
1	1	$z_{11} = \bar{\lambda}_i \bar{\lambda}_j \oplus \lambda_K$	$H\left(L_i^1, L_j^1\right) \oplus \left(L_k^{z_{11}}, z_{11}\right)$

The free-XOR technique The free-XOR technique [4,5] allows XOR gates to be securely computed without any interaction even in the presence of malicious adversaries. The basic idea is to let the circuit garbler keep a global secret Δ and dictate that for every wire i whose 0-label is L_i^0, its 1-label L_i^1 is always defined as $L_i^1 := L_i^0 \oplus \Delta$. Further, for an XOR gate with input wires i, j and output wire k, the garbler will always set $L_k^0 := L_i^0 \oplus L_j^0$. Thus XOR can be securely computed by the evaluator alone through XOR-ing the two input wire-labels it obtained from evaluating previous gates. Hence, the garbled table of the AND gate evolves into

$b_i \oplus \lambda_i$	$b_j \oplus \lambda_j$	$b_k \oplus \lambda_k$	Free-XOR
0	0	$z_{00} = \lambda_i \lambda_j \oplus \lambda_K$	$H\left(L_i^0, L_j^0\right) \oplus \left(L_k^0 \oplus z_{00}\Delta, z_{00}\right)$
0	1	$z_{01} = \lambda_i \bar{\lambda}_j \oplus \lambda_K$	$H\left(L_i^0, L_j^1\right) \oplus \left(L_k^0 \oplus z_{01}\Delta, z_{01}\right)$
1	0	$z_{10} = \bar{\lambda}_i \lambda_j \oplus \lambda_K$	$H\left(L_i^1, L_j^0\right) \oplus \left(L_k^0 \oplus z_{10}\Delta, z_{10}\right)$
1	1	$z_{11} = \bar{\lambda}_i \bar{\lambda}_j \oplus \lambda_K$	$H\left(L_i^1, L_j^1\right) \oplus \left(L_k^0 \oplus z_{11}\Delta, z_{11}\right)$

Authenticated garbling The garbling protocol given earlier can only thwart semihonest adversaries. In the standard *malicious* threat model, however, a malicious circuit generator can put erroneous rows into the garbled table. Based on the values of the permutation bits $\lambda_i, \lambda_j, \lambda_k$ along with the fact of whether the evaluation succeeds, a malicious garbler can learn extra information about the plaintext wire signals involved in the erroneous gates. To thwart such attacks, Wang et al. [6] proposed a seminal technique called *authenticated garbling*.

The basic idea is to hide the permutation bits from any subset of the parties so that in event of a malicious generator corrupting some garbled rows, the adversary should have no clues on which pair of plaintext signals a garbled row is associated with. Meanwhile, authenticated garbling enables the circuit evaluator to locally verify whether a decrypted row was indeed correctly constructed. Therefore a protocol execution will fail or succeed, but in either case, its behavior is independent of any honest party's secret inputs.

Authenticated garbling is by far one of the most practical constant-round actively-secure n-party computation scheme that tolerates any number of corrupted parties. It requires and works with the free-XOR technique. At the high level, authenticated garbling runs in two phases: an offline phase precomputes and stores all abits and aANDs needed later in the protocol, and a function-dependent online phase that generates and evaluates an *authenticated* garbled circuit using the abits and aAND prepared earlier.

The key enabling tool of authenticated garbling is *authenticated AND triples* (aAND), which are precomputed using a separate secure computation protocol. Assume P_1 has a secret value $\Delta_1 \in \{0,1\}^n$. We denote by $[b]^1$ an authenticated bit b of party P_1, which refers to a distributed tuple $(b, M[b], K[b])$ such that $M[b] = K[b] \oplus b\Delta_1$ where P_1 has $(b, M[b])$, and P_2 knows $K(b)$. We call $M[b] \in \{0,1\}^n$ the Message Authentication Code (MAC) of b, and $K[b] \in \{0,1\}^n$ the *verification key* to *b's* MAC. An authenticated bit can be efficiently computed using oblivious transfer: one party plays the OT sender with $K[b]$ and $K[b] \oplus \Delta\, b$ as the two input messages while the other party plays the OT receiver with bit b as the input to learn $M[b]$. An *authenticated AND triple* is just a tuple of six authenticated bits $[a_1]^1, [b_1]^1, [c_1]^1, [a_2]^2, [b_2]^2, [c_2]^2$ such that $(a_1 \oplus a_2)(b_1 \oplus b_2) = c_1 \oplus c_2$. To generate authenticated AND triples, the parties need to run a separate secure protocol in advance. In fact, this protocol, called Π_{aAND}, can dominate the overall cost of this approach of MPC. Wang et al. [6] proposed a Π_{aAND} that works in two high-level steps:

1. Generating leaky-aANDs using Π_{LaAND}. A leaky-aAND triple has the same property as aAND except that a cheating party is able to correctly guess an honest party's first abit output with probability 1/2, at the risk of being caught with probability 1/2.
2. Combine every B randomly chosen leaky-aANDs into a fully secure aAND. The integer B is known as the *bucket size*.

Now we can describe the intuitive ideas behind authenticated garbling in the two-party setting. Let i, j, k be the three wires associated with an AND gate. To hide the permutation bits $\lambda_i, \lambda_j, \lambda_k$, they are divided into XOR-based bit-shares, $[\lambda_i^1]^1, [\lambda_j^1]^1, [\lambda_k^1]^1$ and $[\lambda_i^2]^2, [\lambda_j^2]^2, [\lambda_k^2]^2$, held by P_1 and P_2, respectively, such that $\lambda_i^1 \oplus \lambda_i^2 = \lambda_i, \lambda_j^1 \oplus \lambda_j^2 = \lambda_j, \lambda_k^1 \oplus \lambda_k^2 = \lambda_k$. Now the first question is, how could the circuit generator (call it P_1) produce the first garbled row

$$H\left(L_i^0, L_j^0\right) \oplus \left(L_k^0 \oplus (\lambda_i\lambda_j \oplus \lambda_k)\Delta, (\lambda_i\lambda_j \oplus \lambda_k)\right)$$

without actually knowing $\lambda_i, \lambda_j, \lambda_k$? This is addressed by dividing $(\lambda_i\lambda_j \oplus \lambda_k)\Delta$ into two XOR-shares S^{P_1} and S^{P_2} such that $S^{P_1} \oplus S^{P_2} = (\lambda_i\lambda_j \oplus \lambda_k)\Delta$ where P_1 and P_2 need to locally compute S^{P_1} and S^{P_2}, respectively. However, how can the parties *locally* compute S^{P_1}, S^{P_2} from values that they already know? This is exactly where aANDs come handy: if P_1 and P_2 already have the respective shares of an aAND $\left([a_1]^1 \oplus [a_2]^2\right)\left([b_1]^1 \oplus [b_2]^2\right) = [c_1]^1 \oplus [c_2]^2$ with $a_1 \oplus a_2 = \lambda_i$ and $b_1 \oplus b_2 = \lambda_j$, then P_1, P_2 can learn c_1 and c_2, respectively, with $c_1 \oplus c_2 = \lambda_i, \lambda_j$. Consequently, P_1 can compute $(c_1 \oplus \lambda_k^1)\Delta$ from c_1, λ_k^1 and Δ, all of which P_1 already knows.

Because $\left(c_i \oplus \lambda_k^1\right)\Delta \oplus \left(c_2 \oplus \lambda_k^2\right)\Delta = (\lambda_i\lambda_j \oplus \lambda_k)\Delta$, one would wish P_2 to be able to locally compute $\left(c_2 \oplus \lambda_k^2\right)\Delta$. Unfortunately, P_2 cannot, because it does not know Δ.

To resolve this, one can exploit the fact that P_2 already knows $\mathrm{M}[c_2]$ and $\mathrm{M}\left[\lambda_k^2\right]$, both generated from the same Δ_1 (P_1's global secret for authenticating P_2's bits). Because $\mathrm{K}[b] \oplus \mathrm{M}[b] = b\Delta_1$ for any authenticated bit b, it suffices to require P_1 to set $\Delta = \Delta_1$, and define

$$S^{P_1} \stackrel{\text{def}}{=} \left(c_1 \oplus \lambda_k^1\right)\Delta \oplus \mathrm{K}[c_2] \oplus \mathrm{K}\left[\lambda_k^2\right]$$
$$S^{P_2} \stackrel{\text{def}}{=} \mathrm{M}[c_2] \oplus \mathrm{M}\left[\lambda_k^2\right]$$

so $S^{P_1} \oplus S^{P_2} = (\lambda_i\lambda_j \oplus \lambda_k)\Delta$ for P_1 and P_2 to locally compute S^{P_1} and S^{P_2}, respectively. Finally, to prevent a malicious P_1 from replacing $c_1 \oplus \lambda_k^1$ with an arbitrary bit, authenticated garbling requires P_1 to provide $\mathrm{M}[c_1] \oplus \mathrm{M}\left[\lambda_k^1\right]$ (i.e., the MAC of $c_1 \oplus \lambda_k^1$) so that P_2 can verify the correctness of a garbled row. Moreover, observing that the MACs are XOR-homomorphic and $\lambda_i\overline{\lambda_j} = \lambda_i \oplus \lambda_i\lambda_j$, $\overline{\lambda_i}\lambda_j = \lambda_j \oplus \lambda_i\lambda_j$, and $\overline{\lambda_i}\overline{\lambda_j} = 1 \oplus \lambda_i \oplus \lambda_j \oplus \lambda_i\lambda_j$, the same aAND triple can (and should) be reused to compute the rest three garbled rows in the same AND gate. In summary, P_1 will generate its share of garbled AND, where $H_{b_i,b_j} \stackrel{\text{def}}{=} \mathrm{H}\left(\mathrm{L}_i^{b_i}, \mathrm{L}_j^{b_j}\right)$, $\forall b_i, b_j \in \{0,1\}$, as below

$$\begin{array}{lll}
H_{0,0} \oplus \left(\mathrm{L}_k^0 \oplus \left(c_1 \oplus \lambda_k^1\right)\Delta \oplus \mathrm{K}[c_2] \oplus \mathrm{K}\left[\lambda_i^2\right], \right. & c_1 \oplus \lambda_k^1, & \left. \mathrm{M}[c_1] \oplus \mathrm{M}\left[\lambda_k^1\right]\right) \\
H_{0,1} \oplus \left(\mathrm{L}_k^0 \oplus \left(c_1 \oplus \lambda_k^1 \oplus \lambda_i^1\right)\Delta \oplus \mathrm{K}[c_2] \oplus \mathrm{K}\left[\lambda_k^2\right] \oplus \mathrm{K}\left[\lambda_i^2\right] \right. & c_1 \oplus \lambda_k^1 \oplus \lambda_i^1 & \left. \mathrm{M}[c_1] \oplus \mathrm{M}\left[\lambda_k^1\right] \oplus \mathrm{M}\left[\lambda_i^1\right]\right) \\
H_{1,0} \oplus \left(\mathrm{L}_k^0 \oplus \left(c_1 \oplus \lambda_k^1 \oplus \lambda_j^1\right)\Delta \oplus \mathrm{K}[c_2] \oplus \mathrm{K}\left[\lambda_k^2\right] \oplus \mathrm{K}\left[\lambda_j^2\right], \right. & c_1 \oplus \lambda_k^1 \oplus \lambda_j^1 & \left. \mathrm{M}[c_1] \oplus \mathrm{M}\left[\lambda_k^1\right] \oplus \mathrm{M}\left[\lambda_j^1\right] \right. \\
H_{1,1} \oplus \left(\mathrm{L}_k^0 \oplus \left(c_1 \oplus \lambda_k^1 \oplus \lambda_j^1\right)\Delta \oplus \mathrm{K}[c_2] \oplus \mathrm{K}\left[\lambda_k^2\right] \oplus \mathrm{K}\left[\lambda_i^2\right] \oplus \mathrm{K}\left[\lambda_j^2\right] \oplus \Delta, \right. & c_1 \oplus \lambda_k^1 \oplus \lambda_i^1 \oplus \lambda_j^1 & \left. \mathrm{M}[c_1] \oplus \mathrm{M}\left[\lambda_k^1\right] \oplus \mathrm{M}\left[\lambda_i^1\right] \oplus \mathrm{M}\left[\lambda_j^1\right]\right)
\end{array}$$

while P_2 will generate its share of garbled AND as

$$\begin{array}{lll}
\mathrm{M}[c_2] \oplus \mathrm{M}\left[\lambda_k^2\right], & c_2 \oplus \lambda_k^2, & \mathrm{K}[c_1] \oplus \mathrm{K}\left[\lambda_k^1\right] \\
\mathrm{M}[c_2] \oplus \mathrm{M}\left[\lambda_k^2\right] \oplus \mathrm{M}\left[\lambda_i^2\right], & c_2 \oplus \lambda_k^2 \oplus \lambda_i^2, & \mathrm{K}[c_1] \oplus \mathrm{K}\left[\lambda_k^1\right] \oplus \mathrm{K}\left[\lambda_i^1\right] \\
\mathrm{M}[c_2] \oplus \mathrm{M}\left[\lambda_k^2\right] \oplus \mathrm{M}\left[\lambda_j^2\right], & c_2 \oplus \lambda_k^2 \oplus \lambda_j^2, & \mathrm{K}[c_1] \oplus \mathrm{K}\left[\lambda_k^1\right] \oplus \mathrm{K}\left[\lambda_j^1\right] \\
\mathrm{M}[c_2] \oplus \mathrm{M}\left[\lambda_k^2\right] \oplus \mathrm{M}\left[\lambda_i^2\right] \oplus \mathrm{M}\left[\lambda_j^2\right], & c_2 \oplus \lambda_k^2 \oplus \lambda_i^2 \oplus \lambda_j^2 \oplus 1, & \mathrm{K}[c_1] \oplus \mathrm{K}\left[\lambda_k^1\right] \oplus \mathrm{K}\left[\lambda_i^1\right] \oplus \mathrm{K}\left[\lambda_j^1\right]
\end{array}$$

That is, to garble an AND gate, a circuit generator will invoke the random oracle four times but the circuit evaluator doesn't really need the random oracle. We stress that because the MACs and keys are used in constructing garbled tables, the length of the MACs and keys are essentially a computational security parameter.

The idea of authenticated garbling naturally works for the general nn-party setting. In this setting, every party P_i needs to run $n-1$ OT instances to obtain $n-1$ MACs for every authenticated bit b that it owns. Of course, the extra checking mechanism is needed to ensure that P_i indeed used the same b across all $n-1$ oblivious transfers. In addition, note that in the online circuit garbling and evaluation phase, there will be $n-1$ circuit garbler each of which generates one share of the authenticated garbled circuit but only a single circuit evaluator who receives all $n-1$ garbled circuits, combining them (through XOR-summing all of them together) with its own share of garbled circuit then carrying out circuit evaluation.

Generating authenticated AND triples The parties need to run a separate secure protocol, called Π_{aAND}, in an input-independent preparation phase to generate aAND triples. In fact, the cost of Π_{aAND} dominates that of authenticated garbling described earlier that uses the result of Π_{aAND} Π_{aAND} works in two high-level steps:

1. Generating leaky-aANDs using Π_{LaAND}. A leaky-aAND triple has the same property as aAND except that a cheating party is allowed to guess one of an honest party's abit. A correct guess of a single abit, which occurs with probability 1/2, will remain undetected whereas an incorrect guess will be caught with probability 1/2.
2. Combine every B randomly chosen leaky-aANDs into a fully secure aAND. As a security parameter, B is also known as the *bucket size.*

3.5 Pool-based cut-and-choose

By far, the most efficient implementations of actively secure computation protocols [7–11] are based on the idea of *batched cut-and-choose*. These protocols, however, suffered from scalability issues as they require linear storage (in the length of the computation) because much per-gate/per-copy information needs to be stored *before* the cut-and-choose challenges can be revealed. To overcome this issue, Zhu et al. [12] proposed to maintain a fixed-size pool for keeping the necessary information to do cut-and-choose. Namely, the garbled gates used for checking/evaluation will always be selected from a fixed-size pool and the pool will be refilled immediately after any garbled gate is consumed. Example instantiations of this idea include Pool-JIMU [11] and nanoPI [13], both exhibiting extraordinary scalability advantages. In addition, pool-based cut-and-choose also offers unpaired *long-term* statistical security guarantee, that is, cut-and-choose failures are bounded throughout the *lifetime* of the pool regardless of how many secure computation instances have been executed.

References

[1] Goldreich O, Micali S, Wigderson A. How to play any mental game. In: Proceedings of the nineteenth annual ACM symposium on Theory of computing. ACM; 1987. p. 218–29.

[2] Ben-Or M, Goldwasser S, Wigderson A. Completeness theorems for non-cryptographic fault-tolerant distributed computation. In: STOC; 1988.

[3] Pinkas B, Schneider T, Smart NP, Williams SC. Secure two-party computation is practical. In: ASIACRYPT; 2009.

[4] Beaver D, Micali S, Rogaway P. The round complexity of secure protocols. In: STOC; 1990.

[5] Kolesnikov V, Schneider T. Improved garbled circuit: Free XOR gates and applications. In: ICALP; 2008.

[6] Wang X, Ranellucci S, Katz J. Authenticated garbling and efficient maliciously secure two-party computation. In: ACM CCS; 2017.

[7] Kolesnikov V, Nielsen J, Rosulek M, Ni T, Trifiletti R. Duplo: unifying cut-and-choose for garbled circuits. In: ACM CCS; 2017.

[8] Lindell Y, Riva B. Blazing fast 2pc in the offline/online setting with security for malicious adversaries. In: ACM CCS; 2015.

[9] Nielsen J, Schneider T, Trifiletti R. Constant round maliciously secure 2pc with function-independent preprocessing using lego. NDSS; 2017.

[10] Rindal P, Rosulek M. Faster malicious 2-party secure computation with online/offline dual execution. In: USENIX security symposium; 2016.

[11] Zhu R, Huang Y. JIMU: faster lego-based secure computation using additive homomorphic hashes. In: ASIACRYPT; 2017.

[12] Zhu R, Huang Y, Cassel D. Pool: scalable on-demand secure computation service against malicious adversaries. In: ACM CCS; 2017.

[13] Zhu R, Cassel D, Huang Y, Sabry A. nanoPI: nanoPI: Extreme-scale actively-secure multi-party computation. In: ACM CCS; 2018.

CHAPTER

Game theory for privacy-preserving sharing of genomic data

7

Zhiyu Wan[1], Yevgeniy Vorobeychik[2], Ellen Wright Clayton[3,4,5], Murat Kantarcioglu[6], Bradley Malin[1,7,8]

[1]Department of Electrical Engineering and Computer Science, Vanderbilt University, Nashville, TN, United States; [2]Department of Computer Science and Engineering, Washington University in St. Louis, St. Louis, MO, United States; [3]Center for Biomedical Ethics and Society, Vanderbilt University Medical Center, Nashville, TN, United States; [4]Law School, Vanderbilt University, Nashville, TN, United States; [5]Department of Pediatrics, Vanderbilt University Medical Center, Nashville, TN, United States; [6]Department of Computer Science, University of Texas at Dallas, Richardson, TX, United States; [7]Department of Biomedical Informatics, Vanderbilt University Medical Center, Nashville, TN, United States; [8]Department of Biostatistics, Vanderbilt University Medical Center, Nashville, TN, United States

1. Introduction

Genomic data are increasingly gathered by a wide array of organizations [1], ranging from direct-to-consumer genomic companies [2,3] to clinically focused institutions [4,5]. Meanwhile, emerging large-scale scientific endeavors, such as the All of Us Research Program of US National Institute of Health (NIH) [6], are collecting a variety of data, including genomic and phenomic records on millions of participants. It is anticipated that broad sharing and reuse of these data enable opportunities for scientific discovery and research advancement.

Although broad data sharing is important, the participants whose data have been contributed to these resources often expect that their privacy, and particularly their anonymity, will be preserved [7]. Yet these expectations are eroding due to high-profile demonstrations, conducted over the past decade, that have shown how de-identified genomic data can be traced back to named persons [8,9].

To provide intuition into knowledge derived from genome-based investigations, data custodians have turned toward sharing data in statistically aggregated forms (e.g., sharing only allele frequencies) about the pool of individuals who were in a study or were treated in a clinical setting. The practice of aggregated data sharing began on a large scale in the mid-2000s, with programs like the database of Genotypes and Phenotypes (dbGaP) at the NIH [10], which aimed to standardize and centralize genomic data, making it easier to access. Summary statistics about the allele rates were made publicly accessible over the Internet because it was assumed that the privacy risks for such data were minimal. Yet Homer et al.

Responsible Genomic Data Sharing. https://doi.org/10.1016/B978-0-12-816197-5.00007-3

in Ref. [11] demonstrated that an adversary could apply a statistical-inference attack to detect the presence of a known individual's DNA sequence in a pool of subjects (e.g., a case-control cohort of individuals positively diagnosed with a sexually transmitted disease). Specifically, distances between an individual's sequences to the allele frequencies exhibited by the pool versus some reference population were measured. When the target was deemed to be sufficiently biased toward the pool, the adversary could assign the target with the membership. As a result, the NIH, Wellcome Trust, and other genomic data custodians removed all aggregated data of human genomes from public websites [12]. In Ref. [13], Wang et al. extended Homer's attack by using more aggregated data published by researchers such as *P*-values and coefficients of correlations. Following Homer's work, in Ref. [14], Sankararaman et al. demonstrated the optimal detection power of likelihood ratio statistics and provided quantitative guidelines for the data publisher.

A review of all such demonstrations is beyond the scope of this chapter, but we refer the reader to Naveed et al. [15] and Mittos et al. [16], for review on privacy attacks committed over the past decade, as well as a general framework to systematize the analysis of threats.

In the face of such risks, organizations have adopted various sociotechnical controls. These include legal protections, such as data use agreements (DUAs) that explicitly prohibit re-identification [17]. These also include techniques for privacy-preserving data publishing, such as generalization [18], suppression [14], and obfuscation of potential identifiers [19,20].

Although privacy-preserving data publishing techniques can explicitly trade-off utility and privacy, they fail to account for the behavior of would-be attackers who attempt to violate the privacy of individuals in the shared data. Moreover, the side effect of this line of research, which focuses on worst-case scenarios, has been the increasing promotion and, at times, adoption of data-sharing practices that may unnecessarily impede research [21]. In other words, the community often focuses on what is possible, as opposed to what is probable. Game theory can be invoked to provide a more realistic view on how to measure the risks and recommend protections. A game-theoretic approach offers an alternative that trades off utility and privacy in a way that explicitly accounts for adversarial behavior and capabilities, and can potentially help the data sharers and the policymakers to assess the privacy risk and find the best protection strategy. Game-theoretic approaches have been applied to a variety of security [22–25] and privacy [26–31] problems.

In the remainder of this chapter, we summarize technical approaches that have applied game theory to privacy protection in genomic data sharing in three scenarios. In the first, a data center shares summary statistics to the public and the malicious recipient utilizes a statistical-inference attack to detect the membership of targeted individuals. In the second scenario, participants of a research study choose different protection or sharing choices, while the adversary attempts to infer hidden information about the genomes of their relatives. In the third scenario, a data center shares de-identified individual-level genomic data, while the adversary re-identified those records by linking them to an external dataset. Specifically, in Section 2, we introduce the basic concept of game theory to orient the readers. In Section 3, we review the membership-inference game, which is based on the membership-inference attack targeting genomic data. In Section 4, we introduce a game that is similar to the membership-inference game but with a more powerful attack. In Section 5, we introduce the kinship game, which is based on kinship attacks targeting genomic data. In Section 6, we introduce the re-identification game that is based on a linkage attack for genomic data. Finally, we discuss the differences between these game models and how they relate to the privacy-preserving sharing of genomic data more generally in Section 7 and conclude in Section 8.

2. Background

In this section, we introduce the basics of game theory for those readers who are not familiar with game theory.

Game theory is a branch of applied mathematics that studies the interactions among rational agents and their behaviors [32]. A game G with n players can be defined as a triplet $\langle P, S, U \rangle$, where $P = \langle p_1, \ldots, p_n \rangle$ is the set of players. The set of strategy profiles is $S = S_1 \times \ldots \times S_n$ where S_i is a set of strategies that player i can take. The payoff of a player can be affected by the strategies of all players. Thus, the set of payoff functions is $U(s) = (u_1(s), \ldots, u_n(s))$ where $u_i : S \rightarrow R$.

A strategy profile $s^* \in S$ is a Nash equilibrium if no player can do better by unilaterally changing his or her strategy, that is,

$$u_i\left(s_i^*, s_{-i}^*\right) \geq u_i\left(s_i, s_{-i}^*\right), \forall i \tag{7.1}$$

where s_{-i} is a strategy profile of all players other than player i. When all players adopt a fixed action as their strategy, s^* is a pure-strategy Nash equilibrium. When all players randomly choose an action according to a probability distribution, s^* is a mixed-strategy Nash equilibrium. It has been proved that every finite game has a mixed-strategy Nash equilibrium.

In a two-party Stackelberg game, there are two players: the leader p_1 and the follower p_2. The leader makes the first choice, and its strategy is then observed by the follower.

We assume that the follower then plays the optimal strategy in response to the leader (that is, the *best response*), defined as

$$s_2^*(s_1) = \underset{s_2 \in S_2}{\operatorname{argmax}} u_2(s_1, s_2) \tag{7.2}$$

Using backward induction, a brute force methodology, the leader's best strategy can then be determined:

$$s_1^* = \underset{s_1 \in S_1}{\operatorname{argmax}} u_1\left(s_1, s_2^*(s_1)\right) \tag{7.3}$$

3. Membership-inference game

Sankararaman et al. [14] improved Homer et al.'s membership-inference attack [11] with a log-likelihood ratio (LR) test. Specifically, they estimated the upper bound of the number of single-nucleotide polymorphisms (SNPs) that can be published safely given the size of the pool, the maximal allowable detection power (true-positive rate), and the maximal allowable false-positive rate. Other than the empirical method, they developed an analytical method that can estimate the number of SNPs as a function of the size of the pool and the two bounds with almost no computation effort.

Wan et al. [33] demonstrated how game-theoretic models can be applied to determine the optimal set of genomic data-sharing policies against Sankararaman et al.'s membership-inference attack [14]. The genomic data-sharing process and the membership-inference game are illustrated in Fig. 7.1.

3.1 The game and its solutions

In this model, there are two players: the sharer, who could be an investigator of a study, and the recipient, who requests and accesses the data. A malicious recipient has the potential to infer the presence of identified genomes (that is, DNA linked to the personal names and contact information of the individuals to whom the data corresponds) in the research study. The sharer gains utility from disseminating data, while the recipient benefits by detecting and exploiting the targets. The costs of an attack, if it is enacted, include the cost of accessing the data and the fines for breaching a DUA (that is, if such deviation is detected). The sharer decides the protections to set in place when releasing information about the SNPs and the recipient chooses to enact an attack (provided it is worth its cost). The interaction between the

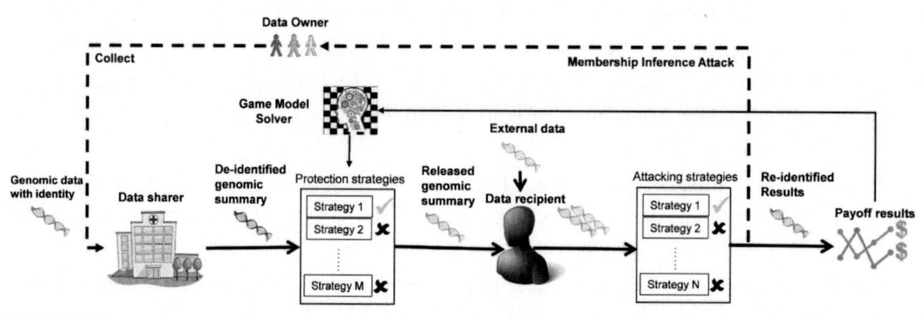

FIGURE 7.1

An illustration of the genomic data-sharing process and the membership-inference game. The genomic data-sharing process follows a series of steps: (1) the data sharer collects identified genomic data from data owners; (2) the sharer releases de-identified genomic summary data to a recipient; and (3) the recipient infers the membership of targets in the study with the assistance of an external dataset and statistical hypothesis tests (e.g., likelihood ratio test). The game components include the following decision points: (1) the recipient selects the optimal attacking strategy given the released summary data and (2) the sharer selects the optimal protection strategy by solving the game model.

two players was naturally modeled using a Stackelberg game model [34]. More specifically, the sharer is a leader who discloses a subset of SNP summary statistics from a study, while the recipient is a follower who determines whether or not to attack each target. The sharer's optimal strategy, which balances the expected utility and privacy risk, can be found as the equilibrium of the game.

To be more precise, the Nash equilibrium of the game can be regarded as the solution to an optimization problem and has the following formal mathematical representation.

Let g be the set of genomic variants (e.g., SNPs) to be shared and a be a set of individuals to be attacked. Then the sharer's optimal strategy is

$$g^* = \underset{g}{\operatorname{argmax}} \overbrace{\left(B_S(g) - \widehat{C}_S(g, a^*(g))\right)}^{\text{Sharer's Estimated Payoff}} \tag{7.4}$$

in which

$$a^*(g) = \underset{a}{\operatorname{argmax}} \overbrace{\left(\widehat{B}_R(g, a) - \widehat{C}_R(a)\right)}^{\text{Recipient's Estimated Payoff}} \tag{7.5}$$

The benefit to the sharer $B_S(g)$ is a function of the sharer's strategy, proportional to the utility of the data:

$$B_S(g) = H \cdot U(g) = H \cdot \sum_{j \in J} w_j g_j, \tag{7.6}$$

where H corresponds to the value of all the data available to the sharer and $U(g)$ is a score in [0, 1] that represents the utility for shared data. J is a set of SNPs and w_j represents the utility weight of SNP j, and $\sum_{j \in J} w_j = 1$. g_j is 1 if the allele frequency for SNP j is shared and 0 otherwise.

The estimated cost for the sharer $\widehat{C}_S(a)$ is a function of the recipient's strategy, proportional to the number of successfully attacked targets:

$$\widehat{C}_S(a) = L_S \cdot \widehat{N}_S(a) = L_S \cdot \sum_{i \in I} a_i \tau_i, \tag{7.7}$$

where L_S is the loss to the sharer per successfully attacked individual. I is a set of targetable individuals. τ_i is the probability that individual i is targeted, and a_i is a binary variable, determined by the recipient's strategy, which is 1 if individual i is attacked and 0 otherwise.

The estimated benefit for the recipient $\widehat{B}_R(g, a)$ is a function of both the sharer's strategy and the recipient's strategy, proportional to the estimated number of successfully attacked targets:

$$\widehat{B}_R(g, a) = G_R \cdot \widehat{N}_R(g, a) = \sum_{i \in I} G_R \cdot p_i l_i(g) a_i \tau_i, \tag{7.8}$$

where G_R is the gain afforded to the recipient per successfully attacked target, p_i is the prior probability that individual i is in the study, and $l_i(g)$ is the likelihood ratio that compares the likelihood that individual i is in the study versus that the individual is in a reference population [14].

The estimated cost to the recipient $\widehat{C}_R(a)$ is a function of the recipient's strategy and is proportional to the number of attacked targets:

$$\widehat{C}_R(a) = c\sum_{i\in I} a_i\tau_i = \sum_{i\in I} c\cdot a_i\tau_i \tag{7.9}$$

where c is the cost to the recipient per attack. As the attacker's decision for each target is independent of the attacker's decisions for other targets, the attacker's best decision for target i is to attack if, and only if, the resulting increment on the recipient's benefit is larger than the resulting increment on the recipient's cost, which is represented as follows:

$$a_i^*(g) = \begin{cases} 1, & |G_R\cdot p_i l_i(g) - c > 0, \\ 0, & |G_R\cdot p_i l_i(g) - c \leq 0, \end{cases} \quad \forall i\in I \tag{7.10}$$

For a large dataset, the strategy space of the sharer can be quite large. As such, the solution to the optimization problem is difficult to discover computationally. Thus, Wan et al. solved the problem [33] using a genetic algorithm, which is an optimization approach inspired by evolutionary processes. In each iteration, the candidates (protection strategies in this case) with higher fitness scores have greater probabilities of being selected.

The experimental evaluation was conducted on a real dataset of 8194 individuals from the Sequence and Phenotype Integration Exchange (SPHINX) program in the Electronic Medical Records and Genomics (eMERGE) project, with 51,826 genetic variants (e.g., SNPs) being collected. The 1000 Genomes Phase 3 resource [35] was used as the reference population.

The results show that the best policy option is realized in a game-theoretic setting that combines an SNP suppression approach with a DUA, in terms of the sharer's payoff. Although the existing "best" option according to Sankararaman et al. [14] improved utility by less than 5%, it traded the utility for an increase of privacy risk by approximately 300% over the game solution. At the same time, it was acknowledged that focusing solely on the payoff may be challenging for institutional review boards to support. As a result, Wan et al. [33] provided a solution to the game that results in no attack being committed (which ensured a recipient will always choose not to attack) by adding an additional set of constraints to the Stackelberg game model:

$$a_i^*(g) = 0, \ \forall i\in I \tag{7.11}$$

To help set a reasonable penalty in the DUA, they further investigated how the results change according to the penalty set in the DUA. As expected, the overall payoff of the sharer increased until the penalty achieves the maximum value the recipient can gain from the attack. However, after a certain point, the growth rate slowed and, eventually became very small. This provides an analytical way for the policymaker to choose a preferred penalty level in the DUA.

In another set of experiments, Wan et al. considered how the prior probability of a target's inclusion in a study influences the results. They set the prior according to four genome sequencing programs: (a) the Precision Medicine Initiative or PMI (which became the All of Us program) [6], which, once its recruitment is complete, will have a prior of 0.003 for 1 million participants out of 318 million US citizens, (b) the Million Veteran Program or MVP [36], which, at the time, had a prior of 0.002 for 400,000 participants out of 21 million US military veterans, (b) the Bio-VU de-identified DNA repository program of the Vanderbilt University Medical Center [37], which has a prior of 0.1 for 200,000 participants out of 2 million individuals with electronic medical records at the institution, and (d) the Rare Diseases Clinical Research Network, with an estimated prior of 0.5. Some notable findings include (a) when the prior probability is relatively small, as in PMI, the difference between the DUA policy and the game policies is negligible, (b) the sharer's payoff is negatively correlated with the prior probability, regardless of the policy, and the game policies are the most robust even when the prior probability increases substantially.

3.2 Limitations of the model

These results demonstrate that blending economic, legal and technical approaches can help strike the right balance between data utility and privacy risk in genomic data sharing and have the potential to revolutionize how policies are designed. However, there are several challenges confronted by this line of research.

First, the settings of parameters are subject to change in different scenarios and are difficult to justify. For example, the value of data is likely dependent upon anticipated usages of the data where the recipient could be an insurer or a blackmailer. In their case study, a successfully attacked target is worth \$360 to the recipient and costs the same amount of money to the data sharer according to an industrial study on the data breach. The cost to access a target is \$60 to the recipient according to the price a typical online data broker charged for a background report. The value of the genomic data is \$45,000 to the sharer according to the grant dollars received for sharing genomic summary statistics. Sensitivity analyses on these important parameters are conducted to justify the settings of parameters in the case study. It turns out that the game policy is always the preferred policy no matter how the values for the parameters vary. Furthermore, the sharer's expected payoff is insensitive to the changes of parameters. This suggests that a rough estimate of parameters might be sufficient to apply the policy in practice.

Second, the game model assumes that the sharer has complete knowledge about the recipient, including the size of the recipient's target set. However, this is unlikely the case in practice. One way to address the uncertainty in the game model would be to use a more complex model, such as a Bayesian game.

Third, the risk assessment is based on the capabilities of an attacker at the present moment, which might improve in the future. Because the protection model expects a one-time release of the summary statistics, no matter how accurate the current models on the parameter valuation or adversarial models are, they would likely change over time. For example, the choice of the optimal solution may be influenced by the accumulation of biological knowledge, the development of computational capability, the advance of algorithmic technology, and the inflation of the currency. Unfortunately, it is nearly impossible to protect the released dataset from future attacks. However, an alternative could be to provision access to the dataset within an accountable environment, such as a registered query-based access system and to release data according to the users' queries. Yet, as the scenario would no longer be a one-time release of data, a more complex game model would be required, such as the sequential game model.

These challenges are not that uncommon for a game-theoretic approach, once being properly addressed, will make the approach more practical in the real world.

Wan et al.'s game-theoretic approach is extendable to other genomic data-sharing scenarios and adversarial models. For instance, by updating the way to compute likelihood ratios, their approach has the potential to address Shringarpure and Bustamante's attack [38], which would be introduced in the next section. Shringarpure and Bustamante's attack, targeting genomic data-sharing beacons, is more powerful than Sankararaman et al.'s attack.

4. Beacon service game

To protect genomic data sharing from membership-inference attacks like Sankararaman et al.'s, the Global Alliance for Genomics and Health created the Beacon service, through which a data holder answers a query about the presence/absence status of a specific allele in a dataset [39]. An example of the query could be "Does the dataset have any genomes with nucleotide A at position 121,212,028 on chromosome 10?" In this way, the information released from the system is restricted to a minimal level but is still useful because, for instance, the observation of a rare allele in multiple datasets might merit further investigation.

Although the Beacon service successfully thwarts Sankararaman et al.'s attack, it is vulnerable to a new attack built upon Sankararaman et al.'s, proposed by Shringarpure and Bustamante (SB) [38], which requires only allele presence information.

To mitigate the privacy risk posed by Beacon services, the Integrating Data for Analysis, Anonymization, and Sharing (iDASH) National Center for Biomedical Computing designed one of the three tracks of their 2016 Genomic Privacy Protection Challenge to explicitly solve this problem [40].

4.1 The game and its solutions

The winning solution was expanded upon and presented by Wan et al. [41], at which time they introduced a generalized framework that is more representative of the real world. Basically, they designed a measurement that helps to accelerate the search process for the best strategy to protect the genomic data against the SB attack, implicitly using a game-theoretic approach. Here we provide additional context on how their method can be explicitly combined with game-theoretic approaches to better protect the genomic data-sharing environment. As in the membership-inference setting, there are two parties in consideration: the defender and the attacker. The attacker is a malicious user launching the SB attack, while the defender is the data holder sharing the genomic data while mitigating the privacy risk of the SB attack. Thus, a Stackelberg game model is again a natural framing, where the leader is the defender and the follower is the attacker. The genomic data-sharing process via the Beacon service and the enhanced membership-inference game are illustrated in Fig. 7.2.

Before jumping into the attacker's strategy space, we describe the attack model. In the SB attack, given a set of Beacon responses, the attacker relies upon a log-likelihood ratio test (LRT) to infer the presence of a targeted genome in the dataset. In the iDASH variation of the SB attack, it is assumed that the attacker knows the alternative allele frequency (AAF) of all single-nucleotide variants (SNVs) in the underlying population.

Formally, let us say that the dataset has m SNVs and n targets. We use $d_i = \{d_{i,1}, \cdots, d_{i,m}\}$ to represent a set of SNVs for target i and $x = \{x_1, \cdots, x_m\}$ to

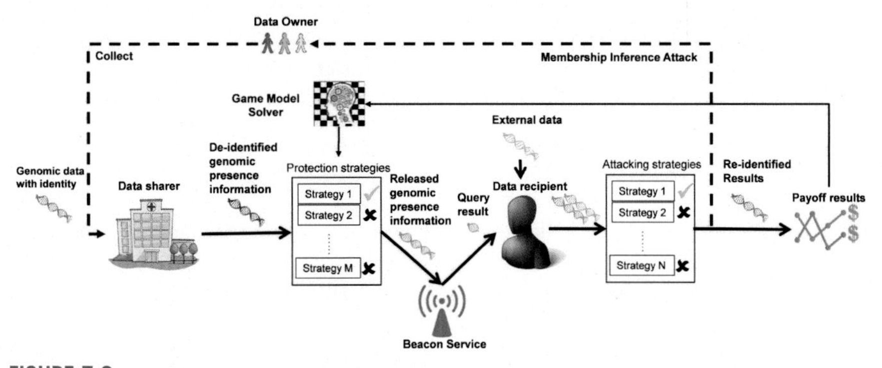

FIGURE 7.2

An illustration of the genomic data-sharing process via the Beacon service and the enhanced membership-inference game. The genomic data-sharing process via the Beacon service follows a series of steps: (1) the data sharer collects identified genomic data from data owners; (2) the sharer releases de-identified genomic presence information to the Beacon server; (3) the recipient queries the Beacon server for the presence information regarding particular genetic variants of the targets; and (4) the recipient infers the membership of targets in the study with the assistance of an external dataset and statistical hypothesis tests (e.g., likelihood ratio test). The game components include the following decision points: (1) the recipient selects the optimal attacking strategy given the released presence information and (2) the sharer selects the optimal protection strategy by solving the game model.

represent a set of Beacon responses. Note that d_{ij} and x_j are binary variables. Specifically, $d_{ij} = 1$ if, and only if, SNV j for target i has at least one alternative allele, $x_j = 1$ if, and only if, SNV j has at least one alternative allele in the beacon. The LRT statistic, as a function of d_i and x, can be represented as follows:

$$L(d_i, x) = \sum_{j=1}^{m} d_{ij} L_j(d_i, x_j), \ \forall i \in I \quad 7.12$$

$$L_j(d_i, x_j) = x_j \log \frac{1 - D_n^j}{1 - \delta D_{n-1}^j} + (1 - x_j) \log \frac{D_n^j}{\delta D_{n-1}^j}, \ \forall i \in I, j \in J \quad 7.13$$

where δ is the sequencing error rate (which is usually set to 10^{-6}), and D_n^j represents the probability that none of the n genomes in the beacon have an alternative allele at SNV j, which is a function of f_j, the AAF of SNV j in the population, as follows:

$$D_n^j = (1 - f_j)^{2n}, \ \forall j \in J \quad 7.14$$

Given this log-likelihood ratio statistic, a threshold is selected, such that only targeted genomes with a test statistic below the threshold are regarded as being in the beacon. In Ref. [41], Wan et al. assumed that the attacker will always select the threshold according to a maximal allowable false-positive rate, which is usually set to 0.05. With this attack model in mind, the data holder can find the best protection strategy by solving an optimization problem. However, a rational attacker will do better by making decisions based on the payoff function instead of adopting a fixed strategy. If we use a Stackelberg model similar to the one mentioned in the last section, according to Eq. (7.10), the attacker best strategy is as follows:

$$a_i^*(x) = \begin{cases} 1, & |L(d_i, x) > \log(c) - \log(b \cdot p_i) = \theta_i \\ 0, & |L(d_i, x) \leq \log(c) - \log(b \cdot p_i) = \theta_i \end{cases}, \ \forall i \in I \quad 7.15$$

where c is the cost per attack, b is the benefit per successful attack, and p_i is the prior probability that target i is in the dataset. In other words, the attacker maximizes his or her payoff so long as the threshold θ_i is set accordingly. Note that the thresholds for different targets are not necessarily the same.

In the context of the iDASH challenge, the only choice the defender can select is to answer "Yes" or "No" to the attacker's query. As it is assumed that the attacker's queries are not accountable (i.e., not in a registered system), the defender needs to select a strategy beforehand, which makes the game a one-shot game. Thus, the number of actions available to the defender is 2^m. In each strategy, for the alternative allele in each position, the defender could choose to answer truthfully or not. If we define the strategy for each SNV as choosing one of these two functions both from the set {"Yes," "No"} (or $\{1, 0\}$) to the set itself, then the number of strategies available to the defender is also 2^m. The first function is disclosing (or D), such that $D(0) = 0$, $D(1) = 1$, and the second function is flipping (or F), such that $F(0) = 1$, $F(1) = 0$.

In a more practical scenario, the defender has a third choice to select for the attacker's query, which is to answer "Not Applicable" (or "NA"). This happens, for instance, when the defender does not have any records that cover the SNV of interest. In this case, the number of available actions to the defender increases to 3^m. If we define the strategy for each SNV as choosing one of these three functions all from the set {"Yes," "No"} (or {1, 0}) to the set {"Yes," "NA," "No"} (or {1, 0.5, 0}), then the number of available strategies to the defender is also 3^m. The first function is disclosing (or D), such that $D(0) = 0$, $D(1) = 1$, and the second function is flipping (or F), such that $F(0) = 1$, $F(1) = 0$ and the third function is masking (or M), such that $M(0) = 0.5$, $M(1) = 0.5$.

Note that the "No" and "NA" answers affect the attack model differently. The additive contributions from the SNV j to the final LRT statistics, according to Eq. (7.13), for answers "Yes," "No," and "NA" are $log\frac{1-D_n^j}{1-\delta D_{n-1}^j}$, $log\frac{D_n^j}{\delta D_{n-1}^j}$, and 0, respectively.

In Ref. [41], Wan et al. only investigated the former case (i.e., the context of the iDASH challenge). Here we consider the latter case, which is likely to be more practical for real-world deployment. The latter case is more complicated than the former one; however, by showing how to solve a slightly different problem, we aim to illustrate how a game-theoretic approach could be applied in a variety of scenarios.

To represent the protection mechanism more formally, let us say $s = \{s_1, \cdots, s_m\}$ corresponds to a strategy for all SNVs. Specifically, $s_j = 1$ if, and only if, the strategy for SNV j is disclosing its original response; $s_j = 0$ if, and only if, the strategy for SNV j is flipping to its opposite response; and $s_j = 0.5$ if, and only if, the strategy for SNV j is masking. The LRT statistic, as a function of d_i, s and x, can be represented as follows:

$$L(d_i, y(s,x)) = \sum_{j\in J} d_{ij} L_j\left(y_j(s_j, x_j)\right), \forall i\in I \quad 7.16$$

$$L_j\left(y_j(s_j, x_j)\right) = y_j(s_j, x_j) log\frac{1 - D_n^j}{1 - \delta D_{n-1}^j} + y_j(s_j, 1 - x_j) log\frac{D_n^j}{\delta D_{n-1}^j}, \forall i\in I, j\in J, \quad 7.17$$

$$y_j(s_j, x_j) = (x_j + s_j - 1)(2s_j - 1), \forall j\in J, \quad 7.18$$

where $y = \{y_1, \cdots, y_m\}$ corresponds to the set of Beacon responses after applying the protection layer for all SNVs in the set J.

With the added protection layer, the attacker's payoff function and best strategy are unchanged. As such, the attacker's best strategy, according to Eq. (7.14), as a function of the Beacon responses after applying the protection layer, is as follows:

$$a_i^*(y) = \begin{cases} 1, & |L(d_i, y) > \theta_i \\ 0, & |L(d_i, y) \le \theta_i \end{cases}, \forall i\in I \quad 7.19$$

If we use a Stackelberg game model similar to the one mentioned in the last section, then the defender's simplified payoff function is as follows:

$$Y(g,a) = w_g \cdot \sum_{j \in J} g_j - w_a \cdot \sum_{i \in I} a_i \quad 7.20$$

where J is a set of SNVs and I is a set of targeted individuals. w_g corresponds to the worth of the correct presence information of each SNV available to the defender, w_a corresponds to the worth of each protected individual to the defender. Specifically, g_j is a binary variable determined by the defender's strategy: $g_j = 1$ if, and only if, SNV j is truthfully disclosed; and a_i is a binary variable representing the attacker's strategy: $a_i = 1$ if, and only if, target i is attacked. Clearly, for each SNV j, g_j is a function of s_j: $g_j = 1$ if, and only if, $s_j = 1$.

$$g_j = (2s_j - 1)s_j, \ \forall j \in J \quad 7.21$$

Note that Eq. (7.20) is just an example of a payoff function. For simplicity, we assume each SNV has the same amount of value to the defender and each target has the same amount of value to the attacker. We also assume the flipping protection mechanism and the masking protection mechanism have the same influence on the defender's utility. In this case, the defender makes the decision according to the monetary incentive. The payoff function could be further relaxed into a weighted sum of the utility and privacy risk measures, $U(g)$ and $R(a)$, as functions of the defender's strategy and the attacker's strategy, respectively:

$$U(g) = \frac{1}{m} \sum_{j \in J} g_j \quad 7.22$$

$$R(a) = \frac{1}{n} \sum_{i \in I} a_i \quad 7.23$$

$$Y(g,a) = W_g \cdot U(g) - W_a \cdot R(a) \quad 7.24$$

where W_g is the weight of the utility measure, and W_a is the weight of the privacy risk measure. Eq. (7.24) will be the same as the previous payoff function, Eq. (7.20), if $W_g = w_g m$ and $W_a = w_a n$.

In summary, the optimization problem can be written as:

$$s^* = \underset{s}{\operatorname{argmax}} W_g \cdot U(g(s)) - W_a \cdot R(a^*)$$

$$s.t. \ a_i^*(s,x) = \begin{cases} 1, & |L(d_i, y(s,x)) > \theta_i \\ 0, & |L(d_i, y(s,x)) \le \theta_i \end{cases}, \ \forall i \in I \quad 7.25$$

$$\theta_i = \log(c) - \log(b \cdot p_i), \ \forall i \in I$$

A typical payoff to the defender is dependent upon both the defender's and the attacker's strategies. It also depends upon both the utility and privacy measures of the defender's strategy to be effective. However, it does not have to be a linear combination of the utility and privacy measures. In the iDASH Challenge, the effectiveness of a strategy was defined as the number of answers that can be correctly served before a certain portion ($r = 0.6$) of targeted individuals' presence is successfully detected. In this case, the optimization problem can be written as follows:

$$s^* = \underset{s}{\operatorname{argmax}} j^*$$

$$s.t. \ \frac{1}{n}\sum_{i \in I} a^*_{ij}(s,x) \le r, \ \forall j \le j^*$$

$$a^*_{ij}(s,x) = \begin{cases} 1, & \sum_{j=1}^{k} d_{ij} L_j(y_j(s_j,x_j)) > \theta_i \\ 0, & \sum_{j=1}^{k} d_{ij} L_j(y_j(s_j,x_j)) \le \theta_i \end{cases}, \ \forall i \in I, k \in J \quad 7.26$$

$$L_j(y_j(s_j,x_j)) = y_j(s_j,x_j) log\frac{1 - D^j_n}{1 - \delta D^j_{n-1}} + y_j(s_j, 1 - x_j) log\frac{D^j_n}{\delta D^j_{n-1}}, \ \forall i \in I, j \in J$$

$$\theta_i = \log(c) - \log(b \cdot p_i), \ \forall i \in I$$

These are not optimization problems with direct analytical solutions. However, it is straightforward to solve these using empirical methods. Basically, to do so, one can initiate a search from a strategy and iteratively uncover better strategies from the neighborhood of the current strategy until no better strategy can be found. Moreover, there are some principles one can exploit to accelerate the search process. For instance, as the attacker does not make the decision according to the false-positive rate, the measurement on SNVs for discriminative power, based on the reference population, as proposed in Ref. [41] is no longer useful. To find a good starting point, we can flip or mask the top k SNVs with the greatest detection power.

We define the detection power of SNV j as the additional LRT statistic contributed by SNV j. As we assume the flipping and masking mechanisms have the same influence on the defender's utility, the mechanism with lower detection power will be preferred for SNV j.

$$L_j(y_j(0,x_j)) = y_j(0,x_j) log\frac{1 - D^j_n}{1 - \delta D^j_{n-1}} + y_j(0, 1 - x_j) log\frac{D^j_n}{\delta D^j_{n-1}}, \ \forall j \in J \quad 7.27$$

$$L_j(y_j(0.5,x_j)) = y_j(0.5,x_j) log\frac{1 - D^j_n}{1 - \delta D^j_{n-1}} + y_j(0.5, 1 - x_j) log\frac{D^j_n}{\delta D^j_{n-1}}, \ \forall j \in J \quad 7.28$$

Given the original response x_j, the preferred mechanism s_j is determined (i.e., $s_j = 0$ or $s_j = 1$). In the case where SNV j has at least one alternative allele in the beacon (i.e., $x_j = 1$):

$$L_j(y_j(0,1)) = log\frac{D^j_n}{\delta D^j_{n-1}}, \ \forall j \in J \quad 7.29$$

$$L_j(y_j(0.5,1)) = 0, \ \forall j \in J \quad 7.30$$

The flipping mechanism is preferred if, and only if, $logD^j_n < log\delta D^j_{n-1}$ (or $f_j > 1 - \sqrt{\delta}$).

In the case where SNV j has no alternative alleles in the beacon (i.e., $x_j = 0$):

$$L_j(y_j(0,0)) = log\frac{1 - D^j_n}{1 - \delta D^j_{n-1}}, \forall j \in J \tag{7.31}$$

$$L_j(y_j(0.5,0)) = 0, \forall j \in J \tag{7.32}$$

Here, the flipping mechanism is preferred if, and only if, $log\left(1 - D^j_n\right) < log\left(1 - \delta D^j_{n-1}\right)$ (or $f_j < 1 - \sqrt{\delta}$). When no protection mechanism is applied, we have the following:

$$L_j(y_j(1,1)) = log\frac{1 - D^j_n}{1 - \delta D^j_{n-1}}, \forall j \in J \tag{7.33}$$

$$L_j(y_j(1,0)) = log\frac{D^j_n}{\delta D^j_{n-1}}, \forall j \in J \tag{7.34}$$

We define the detection power of SNV j as follows:

$$\Lambda_j(y_j(s_j, x_j)) = \sum_{i \in I} d_{ij} L_j(y_j(s_j, x_j)), \forall j \in J$$

We define the difference of detection power as the maximal difference of LRT statistics between a truthful answer and an untruthful answer for each SNV:

$$D_j(x_j) = \max\left(\Lambda_j(y_j(0,x_j)) - \Lambda_j(y_j(1,x_j)), \Lambda_j(y_j(0,x_j)) - \Lambda_j(y_j(0.5,x_j))\right), \forall j \in J \tag{7.35}$$

In the case of $x_j = 1$ and $f_j \le 1 - \sqrt{\delta}$, this can be simplified to be

$$D_j = log\frac{1 - D^j_n}{1 - \delta D^j_{n-1}}, \quad \forall j \in J \tag{7.36}$$

This function decreases monotonically when f_j, the AAF of SNV j in the population, increases. As a result, SNVs with several lowest AAFs are masked in the starting search point. The defender does not need to prepare strategies for the case of $x_j = 0$ because the targeted individuals are not in the dataset when the answer is "No," and by flipping or masking this response, the privacy measure will remain constant while the utility will decrease. The defender does not need to prepare strategies for the case of $f_j > 1 - \sqrt{\delta}$, because when δ is a very small number, this case will not exist.

Five alternative protection methods are introduced for comparison purposes. The first is called the truthful method, where the defender simply responds to all queries truthfully. The second is called the baseline method, where the defender flips a fixed percent of the SNVs with the lowest AAF in the underlying population. The method is used in the iDASH challenge as a lower bound. It is also the starting point in our new search protocol. The third is called the greedy accountable method, which assumes the users' queries are accountable, such that upon receiving the next SNV query, the defender flip or mask the answer if, and only if, the power of the resulting LLR test is the best. The fourth method is called the random flip method, which was proposed by Raisaro et al. [42]. In this method, the defender flips a portion of SNVs that exhibit unique alleles in the beacon. The fifth method is called the query budget method, which was also proposed by Raisaro et al. [42]. In this method, a privacy budget is assigned to each individual in the beacon. Each time a record contains the queried allele, the budget for that user is reduced by a certain amount. Once a record's budget is exhausted, their genome will no longer contribute to responses provided by the beacon. However, this method requires the users' queries to be accountable.

The experimental evaluation in Ref. [41] is conducted with a real dataset based on the first 400,000 SNVs in Chromosome 10. The pool is composed of 250 individuals randomly selected from the 2504 individuals in Phase 3 of the 1000 Genomes Project [43]. The reference includes 250 individuals randomly selected from the remaining individuals in Phase 3 of the project.

The experimental results showed that the proposed method outperforms all posited baseline and state-of-the-art methods (that were applicable to real-world scenarios) regardless of how key parameters that drive the attack (e.g., the effectiveness measure, the number of records behind the beacon, and the attacker's estimate of allele frequency) vary. In most scenarios, the advantages of the proposed method over other alternative methods are substantial. The effectiveness of our proposed method is larger than all of the alternative methods when the value for the k parameter is smaller than five.

It is anticipated that using a game-theoretic approach leads to a better payoff, which represents a better balance between the data utility and privacy risk. In the method that implicitly uses the game-theoretic approach, the privacy risk was overestimated by not considering the attacker's costs, which include the cost of access and the cost of penalty. However, introducing the cost of penalty to the model needs two prerequisites. First, it requires each user to sign a data use agreement before querying a Beacon service. Second, it requires that the malicious users can be detected as attempting to re-identify records and could be pursued as violators of a contract and penalized for liquidated damages. Neither of these two perquisites is an easy task. As pointed out by Craig and colleagues [44], the prior probability that a targeted individual is actually in the pool is likely to be much smaller than 50% as was the case for the iDASH challenge. By not considering the prior probability, the privacy risk was overestimated.

4.2 Limitations of the model

There are several limitations to this protection method that should be made evident. First, the approach regards each SNV as having the same utility. This is critical to recognize because all of the defender's strategies trade between utility and privacy. The proposed method first masks answers for the SNVs that have the largest detection power. Yet, if these SNVs happen to have a more severe influence on utility than other SNVs, then the best strategy for the defender will likely change. Second, the parameter k in the method determines the starting point of the search for local optimal strategy. A well-specified value of k increases the probability that the local optimal strategy is also globally optimal. The best choice for k is dependent upon a number of factors, including (1) the size of the pool, (2) the number of SNVs, (3) the maximal allowable false-positive rate, (4) the specific data in the pool, and (5) the target set. In practice, when the defender needs to determine k, he or she could simulate an attacker and then select the best choice empirically.

5. Kinship game

The genomic data of family members' are correlated. Based on this fact, Humbert et al. [45] proposed a reconstruction attack that is based on graphical models and belief propagation. The adversary wants to reconstruct actual genomic sequences of a family from observed genomic sequences of them. In other words, the adversary can infer the target's genome sequence from his or her relatives' sequences that are publicly available. The background knowledge of the adversary includes familial relationships, linkage disequilibrium, and minor allele frequencies.

5.1 The game and its solutions

However, members of the same family may have different opinions about how to protect genomic data. An individual's decision on how to share genomic data will not only affect their own genomic privacy but also their family members'. In Ref. [46], Humbert et al. used game-theoretic approaches to study the interactions between members of a family regarding sharing genomic data. The interdependent genomic data-sharing process and the kinship game are illustrated in Fig. 7.3. In this environment, there is no data holder that collects data and makes the decision for a pool of individuals. Instead, each user (i.e., data owner) makes a decision on his or her own. The decision to share genomic data affects their own utility and privacy but also other's privacy. Two games were developed: (1) a storage-security game and (2) a disclosure game. In each game, there are two settings. In a rational setting, the player considers his or her interest only; whereas, in an altruistic setting, the player considers the interest of the whole family (i.e., the summation of interests of all members in the family). It should be recognized that in this environment, the adversary is modeled as the probability of a successful attack as opposed to a strategic agent. The results of the investigation illustrate that altruism does not always lead to a better overall outcome.

In both games, there are N players, each of whom is a member of a family and has had their genome sequenced. In the first game, a user stores his or her genomic data on a personal device and may invest to secure it. In the second game, a user may

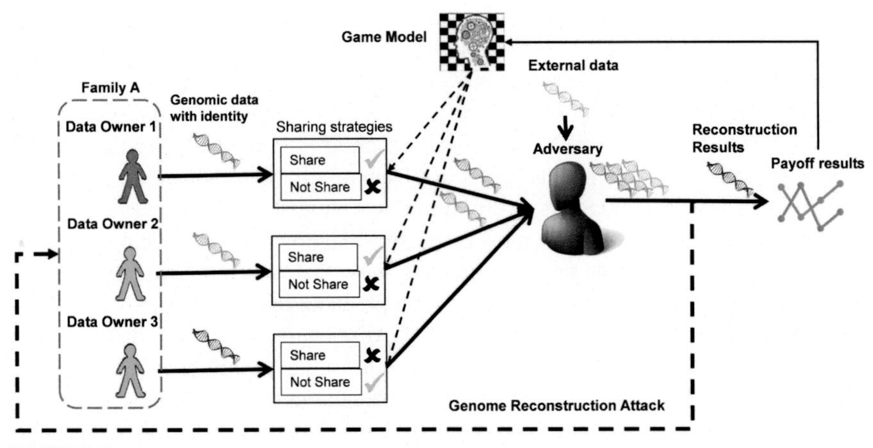

FIGURE 7.3

An illustration of the interdependent genomic data-sharing process and the kinship game. The genomic data-sharing process follows a series of steps: (1) the data owners release genomic data to an adversary and (2) the adversary reconstructs the genomes of the data owner's relatives. The game components include the following decision points: each data owner selects the optimal protection strategy (i.e., invest or not in the storage-security game and disclose or not in the disclosure game).

publicly share his or her genomes on websites such as OpenSNP.org. Thus, the strategy for each player can be represented as a binary variable (x_i or d_i). In the storage-security game, $x_i = 1$ represents an investment in securing his or her own device, while $x_i = 0$ represents no investment. In the disclosure game, $d_i = 1$ represents complete disclosure and $d_i = 0$ represents no disclosure. The payoff for a player is equal to the difference between their benefit and cost. The benefits of a player include the benefit of getting his or her genome sequenced for clinical purposes and the benefit of disclosing his or her genomic data online for secondary usage. The costs for a player include the cost of investment in protection and the loss sustained due to a breach of genomic privacy (i.e., a disclosure of genomic information that they did not disclose by themselves).

The factor, that is interdependent among the players, is the privacy loss. The privacy loss for a player under a strategy profile (i.e., all players' strategies) is dependent upon the estimation error, in terms of the adversary's estimation upon this player's SNP values (i.e., the difference between the estimation error when no SNPs are observed and the estimation error under current strategy profile). In the storage-security game, it is assumed that the probability of a successful breach is 0 if $x_i = 1$, and is between 0 and 1 otherwise. By contrast, in the disclosure game, the probability of a successful breach is 0 if $d_i = 0$ and 1 otherwise. In the altruistic setting, a player maximizes the weighted sum of all relatives' payoffs (including him/herself), in which the weight of a player is equal to a fixed familial factor (between 0 and 1) to the power of degree of kinship.

The two-player game (i.e., $N = 2$) assumes the cost for each player is the same. The solution of a game (i.e., the Nash equilibria) depends on the valuation of the cost and the probability of a successful breach and can be obtained analytically. Each SNP is assumed to be independent of all others. Given the SNP values of parents, the probability distribution of the SNP's value for each child is computed. For the storage game, a genomic dataset of nine relatives in a Utah family (4 grandparents, 2 parents, and 3 children) with 82,000 SNPs on chromosome 1 was used in the experiments. The experimental results show that, in most cases, two players follow identical strategies. For the disclosure game, the analytical results show that, when the discrepancy between the benefits perceived by the players is high enough, these players will follow opposite strategies. In the altruistic setting of the disclosure game, by considering the other player's utility, the analytical results show that a player will choose to disclose his or her genome for a value of benefit higher than that in the selfish setting. In other words, altruism will help reduce the privacy loss caused by the other player. The game model also analytically finds the strategy profile that maximizes the social welfare (i.e., the summation of all players' payoffs) and computes the price of anarchy, which measures how the social welfare decreases due to selfishness. It was also pointed out that, contrary to intuition, altruism in a family does not necessarily lead to higher social welfare. The experiments conducted on two members in the Utah family with 82,000 SNPs confirm the analytic results.

In the N-player disclosure game, with more than two players (i.e., $N > 2$), the payoff functions and game solutions can no longer be expressed in a closed-form. As such, the multiagent influence diagram model, as an extension of the Bayesian network with the additional decision and utility variables, is introduced to represent the game. Most importantly, the game model assumes that the players make decisions sequentially with perfect information. The dependencies of players' decisions could be represented as a directed acyclic graph, from which a topological order is derived. Backward induction was used to solve the sequential game. Experiments are conducted on the nine members of the Utah family with 1000 random chosen SNPs of chromosome 1. The results show that misaligned incentives have a negative impact on social welfare. However, the altruistic setting is not considered in the N-player game.

There are several contributions to this game model. First, it models the problem of kinship privacy in a game-theoretic setting. In doing so, it highlights the fact that the decision of sharing one's own genome can affect all of his/her relative's privacy. Second, it models the situation where each participant in a study has the ability to make decisions. In addition, it models multiple players in a sequential manner. Furthermore, it considers how social welfare is affected by the cooperative and altruistic settings.

5.2 Limitations of the model

At the same time, there are several limitations to be aware of. First, the adversary is not modeled as a player in the game model. Summarizing the adversary's influence on the game as a probability function makes the adversarial model unclear and oversimplified. Second, the strategy space of each player in the considered games is too small. In all games considered, each player only has two choices. In the storage-security game, a player chooses to invest in security or not. In the disclosure game, a player chooses to disclose his or her genomes or not. In a user-centric data-sharing model, a user should have the sharing options with more granularity in terms of whom to share with and which portion of the data to disclose. Third, the parameter settings are not based on real-world settings. Although real-world data are used in experiments to verify the analytic solutions, the parameters such as benefits and costs are still expressed as metrics between 0 and 1, irrelevant to real-world scenarios. For the same reason, it is not clear which variable can be controlled to mitigate privacy risk and optimize social welfare.

6. Re-identification game

In open-access environments such as OpenSNP [47] and the Personal Genome Project [48], participants publish their genomic sequences (e.g., SNPs) to the platform with demographic information for the purpose of facilitating research studies. In Ref. [8], Sweeney et al. inferred the surnames from the misconfigured file names, which are used to verify the correctness of their linkage attacks upon quasi-identifiers (e.g., demographic attributes). Privacy risks exist in controlled-access environments as well, where de-identified genomic datasets from research or healthcare entities are shared with third parties for secondary investigations. The recipients of these datasets could be malicious and want to re-identify the genome records by committing linkage attacks.

6.1 The game and its solutions

In Ref. [49], Wan et al. utilized a game-theoretic framework to analyze the re-identification risks. The health data-sharing process and the re-identification game are illustrated in Fig. 7.4. The framework was illustrated through a case study in genomic data sharing, where the publishers are biomedical researchers who are disseminating research datasets. The NIH requires researchers who are granted funding for genome-based datasets to deposit and share data through online repositories, such as dbGaP. At the same time, the publishers also have an incentive to protect the identities of the individuals who participated in the original research. A malicious recipient of the genomic data would try to re-identify the records in the genetic dataset by linking it to an external dataset and benefit in various ways, such as contacting the participants for marketing purposes, blackmailing the participants using inferred sensitive attributes, or publishing the results to gain academic rewards. A two-player Stackelberg game model was used to model this scenario, in which the leader is the publisher and the follower is the malicious recipient.

The probability that the recipient will successfully re-identify a record is dependent upon the quality (i.e., preciseness and completeness) of the information released by the publisher. A typical technique to sanitize a dataset for anonymization

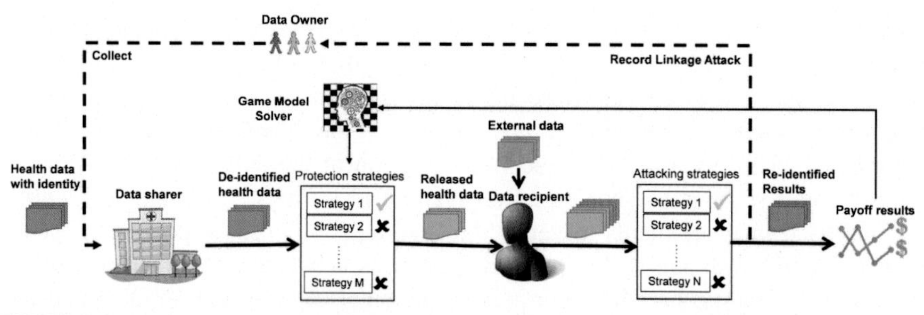

FIGURE 7.4

An illustration of the health data-sharing process and the re-identification game. The health data-sharing process follows a series of steps: (1) the data sharer collects identified health data (including genomic data) from data owners; (2) the sharer releases de-identified health data to a recipient; and (3) the recipient re-identifies targets in the study by linking them to an external dataset upon a set of demographic quasi-identifiers (e.g., age, race, gender). The game components include the following decision points: (1) the recipient selects the optimal attacking strategy given the released data and (2) the sharer selects the optimal protection strategy by solving the game model.

purposes is the generalization technique. For example, the actual value of an age attribute could be generalized to several hierarchical levels: (20–25), (20–29), (0–50), or * (i.e., any age). Thus, the size of the publisher's strategy space is dependent upon the number of quasi-identifiers (i.e., linking attributes) and generalization levels for each quasi-identifier.

Given a record released at a certain generalization level, the recipient has two options: either attempt the re-identification attack or not. A successful re-identification of a record results in loss to the publisher and gain to the attacker. The recipient will only spend a fixed cost when the attack is attempted. The publisher will always gain a fixed payout for a record, depending on its generalization level.

The publisher's benefit is proportional to the amount of funding provided to investigators via NIH grants. It is also affected by the information loss of the released records. An entropy-based metric is defined to measure information loss. The publisher's cost, as a result of privacy loss, is assumed to be proportional to the number of re-identified records and dependent upon the fine paid to US Department of Health and Human Services for privacy violation.

The recipient's benefit is assumed to be the same as the publisher's privacy loss. The probability of successfully re-identifying a record is inversely proportional to the size of the group in the external dataset that matches the record. The cost of the recipient is the fixed price for accessing a record in the external resource.

The solution of the game (i.e., the Nash equilibrium) could be found using two search algorithms: (1) backward induction, an exhaustive search algorithm that is more suitable for a relatively small search space and (2) lattice-based algorithm, a heuristic-driven approach that suits a relatively large search space by pruning nodes.

The experiments are based on a real-world dataset that consists of 32,561 US Census records with demographic attributes including age, race, gender, and 5-digit ZIP codes. The generalization levels for each of them are 6, 4, 2, and 6, respectively.

The results show that the publisher's optimal strategy for sharing the genomic dataset with demographic attributes while protecting privacy could always be found using the game-theoretic approach. Most importantly, it is actually possible to achieve zero risks. The zero-risk solution shares nearly as much data as the optimal one. It also shares much more data than would be shared under the HIPAA Safe Harbor policy. A publishing strategy that compliant with Safe Harbor yields a higher payoff to the publisher due to the reduced privacy risk. These findings are robust to order-of-magnitude changes in parameters such as gains and losses to the publisher and the recipient. In terms of computational performance, the lattice-based algorithm runs much faster than the baseline backward induction algorithm and returns almost the same average payoff for the publisher (99.5% of the solutions are optimal).

6.2 Limitations of the model

There are several limitations of this game model that should be noted. First, the case study assumed a single source of external data. Thus, the results could potentially change if other related external data can be utilized. Second, the study imposes a fixed generalization hierarchy for all records, thereby significantly limiting the granularity of the publisher's decision space. In addition, the publisher could expand his or her strategy space by utilizing other protection strategies such as adding noise to published data. Furthermore, the model assumes a single adversary (data recipient). It would be desirable to generalize the model to capture the scenario of multiple data recipients, the uncertainty about the payoffs and information of data recipients, which would lead to a multi-follower Bayesian Stackelberg game. When multiple adversaries are involved, cooperation and competition among them and the public good should be also considered.

7. Discussion

Although the aforementioned investigations all developed game-theoretic approaches to protect genomic privacy, they have substantial differences. First, the membership-inference game, the Beacon service game, and the re-identification game all have two players: the data protector (i.e., the data publisher) and the adversary (i.e., the data recipient). The adversary in the kinship game is not modeled as a player in the game. As a result, no clear adversarial model is mentioned. Instead, members of a family who participated in the study are modeled as players that make strategic decisions without coordination. Second, in the kinship game, the protector's decisions for all SNPs are the same, while in other games, the protector's decision for each SNP can be different. Thus, the kinship game has a much smaller strategy space to consider, such that searching for the Nash equilibrium is the least challenging. Third, only the kinship game considers the cooperative situation, while all other games are non-cooperative games. Thus, the social welfare and the price of anarchy could be computed. More details regarding the differences among these four games are summarized in Table 7.1.

Table 7.1 Summarized comparison of four genomic data privacy games.

	Membership inference	Beacon service	Kinship	Re-identification
Game model	Stackelberg	Stackelberg	Normal-form game and sequential	Stackelberg
Number of players	2	2	More than 2	2
Released data type	Genomic statistics	Genomic statistics	Genomic data	Genomic data and demographic data
Released data	Allele frequencies	Allele presences	Individual SNPs with pedigree	Individual SNPs with demographics
Privacy threat	Attribute (membership) disclosure	Attribute (membership) disclosure	Genotype reconstruction	Identity disclosure
Attack method	Likelihood ratio test	Likelihood ratio test	N/A	Linkage attack
Adversary's background knowledge	Targets' SNPs with identity and allele frequencies in the population	Targets' SNVs with identity and allele frequencies in the population	N/A	Voter registration lists
Protection actions	Masking	Flipping or masking	Suppression	Generalization
Open source tool	Yes	No	No	No
Dataset	eMERGE SPHINX, 1000 Genomes phase 3	1000 Genomes phase 3	CEPH/Utah family	Adult (US census)
Dataset size	51,826 SNPs for 10,698 individuals	400,000 SNVs for 500 individuals	82,000 SNPs for 9 relatives	4 attributes for 32,561 records

Abbreviations: CEPH, *Centre d'Etude du Polymorphism Humain;* eMERGE, *Electronic Medical Records and Genomics;* SNP, *single-nucleotide polymorphism;* SNV, *single-nucleotide variant;* SPHINX, *Sequence and Phenotype Integration Exchange.*

8. Conclusions

By using game-theoretic approaches, the aforementioned works have modeled the interplay between players with different incentives and have predicted their behaviors at equilibrium. First, all of the genomic privacy game models study a two-player game. In the kinship game, study participants from the same family are players in the game, whereas, in all the other games, the data publisher and the data recipient are players in the game. The proposed models are believed to be able to help the decision-makers choose how to optimally protect the privacy of their genomic data while still helping medical research benefiting from the merits of genomics. In future work, the adversary could be modeled as a player in the kinship game and all the other games could consider the N-player scenario.

Acknowledgments

This research was funded, in part, by the following grants from the National Institutes of Health: U01HG008701, RM1HG009034, R01HG006844, and R01LM009989.

References

[1] Stephens ZD, Lee SY, Faghri F, Campbell RH, Zhai C, Efron MJ, Iyer R, Schatz MC, Sinha S, Robinson GE. Big data: astronomical or genomical? PLoS Biology 2015; 13(7):e1002195.

[2] Phillips AM. Only a click away—DTC genetics for ancestry, health, love… and more: a view of the business and regulatory landscape. Applied and Translational Genomics 2016;8:16–22.

[3] Hazel JW, Slobogin C. Who knows what, and when: a survey of the privacy policies proffered by US direct-to-consumer genetic testing companies. Cornell Journal of Law and Public Policy 2018;28:35.

[4] Rehm HL. Disease-targeted sequencing: a cornerstone in the clinic. Nature Reviews Genetics 2013;14(4):295.

[5] Taber KAJ, Dickinson BD, Wilson M. The promise and challenges of next-generation genome sequencing for clinical care. JAMA Internal Medicine 2014;174(2):275–80.

[6] Denny JC, Rutter JL, Goldstein DB, Philippakis A, Smoller JW, Jenkins G, Dishman E. The "all of us" research program. New England Journal of Medicine 2019;381(7): 668–76.

[7] Trinidad SB, Fullerton SM, Bares JM, Jarvik GP, Larson EB, Burke W. Genomic research and wide data sharing: views of prospective participants. Genetics in Medicine 2010;12(8):485–95.

[8] Sweeney L, Abu A, Winn J. Identifying participants in the personal genome project by name (a re-identification experiment). arXiv 2013. 1304.7605.

[9] Erlich Y, Narayanan A. Routes for breaching and protecting genetic privacy. Nature Reviews Genetics 2014;15(6):409–21.

[10] Mailman MD, Feolo M, Jin Y, Kimura M, Tryka K, Bagoutdinov R, Hao L, et al. The NCBI dbGaP database of genotypes and phenotypes. Nature Genetics 2007;39(10): 1181.

[11] Homer N, Szelinger S, Redman M, Duggan D, Tembe W, Muehling J, Pearson JV, Stephan DA, Nelson SF, Craig DW. Resolving individuals contributing trace amounts of DNA to highly complex mixtures using high-density SNP genotyping microarrays. PLoS Genetics 2008;4(8):e1000167.

[12] Zerhouni EA, Nabel EG. Protecting aggregate genomic data. Science 2008;322(5898): 44.

[13] Wang R, Li YF, Wang X, Tang H, Zhou X. Learning your identity and disease from research papers: information leaks in genome wide association study. In: Proceedings of the 16th ACM conference on computer and communications security, 9–13 November 2009, Chicago, IL. New York, NY, USA: ACM; 2009. p. 534–44.

[14] Sankararaman S, Obozinski G, Jordan MI, Halperin E. Genomic privacy and limits of individual detection in a pool. Nature Genetics 2009;41(9):965.

[15] Naveed M, Ayday E, Clayton EW, Fellay J, Gunter CA, Hubaux J-P, Malin BA, Wang X. Privacy in the genomic era. ACM Computing Surveys (CSUR) 2015;48(1):6.

[16] Mittos A, Malin B, de Cristofaro E. Systematizing genome privacy research: a privacy-enhancing technologies perspective. In: Proceedings on privacy enhancing technologies, vol. 1; 2019. p. 87–107.

[17] Paltoo DN, Rodriguez LL, Feolo M, Gillanders E, Ramos EM, Rutter JL, Sherry S, et al. Data use under the NIH GWAS data sharing policy and future directions. Nature Genetics 2014;46(9):934.

[18] Malin BA. Protecting genomic sequence anonymity with generalization lattices. Methods of Information in Medicine 2005;44(5):687–92.

[19] Jiang X, Zhao Y, Wang X, Malin B, Wang S, Ohno-Machado L, Tang H. A community assessment of privacy preserving techniques for human genomes. BMC Medical Informatics and Decision Making 2014;14(Suppl. 1):S1.

[20] Simmons S, Sahinalp C, Berger B. Enabling privacy-preserving GWASs in heterogeneous human populations. Cell Systems 2016;3(1):54–61.

[21] Barth-Jones D, El Emam K, Bambauer J, Cavoukian A, Malin B. Assessing data intrusion threats. Science 2015;348(6231):194–5.

[22] Pita J, Tambe M, Kiekintveld C, Cullen S, Steigerwald E. GUARDS: game theoretic security allocation on a national scale. In: Proceedings on the 10th international conference on autonomous agents and multiagent systems 1, 2–6 May 2011, Taipei, Taiwan. Richland, SC, USA: International Foundation for Autonomous Agents and Multiagent Systems; 2011. p. 37–44.

[23] Yin Z, Jiang AX, Tambe M, Kiekintveld C, Leyton-Brown K, Sandholm T, Sullivan JP. TRUSTS: scheduling randomized patrols for fare inspection in transit systems using game theory. AI Magazine 2012;33(4):59.

[24] Tambe M, Jain M, Pita JA, Jiang AX. Game theory for security: key algorithmic principles, deployed systems, lessons learned. In: Proceedings of the 2012 50th annual allerton conference on communication, control, and computing (Allerton), 1–5 October 2012, Monticello, IL. Piscataway, NJ, USA: IEEE; 2012. p. 1822–9.

[25] An B, Brown M, Vorobeychik Y, Tambe M. Security games with surveillance cost and optimal timing of attack execution. In: Proceedings of the 2013 international conference on autonomous agents and multi-agent systems, 6–10 May 2013, St. Paul, MN.

Richland, SC, USA: International Foundation for Autonomous Agents and Multiagent Systems; 2013. p. 223–30.

[26] Nix R, Kantarcioglu M. Incentive compatible privacy-preserving distributed classification. IEEE Transactions on Dependable and Secure Computing 2011;9(4): 451–62.

[27] Blocki J, Christin N, Datta A, Procaccia AD, Sinha A. Audit games. In: Proceedings of the twenty-third international joint conference on artificial intelligence, 3–9 August 2013, Beijing, China. Menlo Park, CA, USA: AAAI Press; 2013. p. 41–7.

[28] Blocki J, Christin N, Datta A, Procaccia AD, Sinha A. Audit games with multiple defender resources. In: Proceedings of the twenty-ninth AAAI conference on artificial intelligence, 25–30 January 2015, Austin, TX. Menlo Park, CA, USA: AAAI Press; 2015. p. 791–7.

[29] Li M, Carrell D, Aberdeen J, Hirschman L, Kirby J, Li B, Vorobeychik Y, Malin BA. Optimizing annotation resources for natural language de-identification via a game theoretic framework. Journal of Biomedical Informatics 2016;61:97–109.

[30] Oh SJ, Fritz M, Schiele B. Adversarial image perturbation for privacy protection a game theory perspective. In: Proceedings of the 2017 IEEE international conference on computer vision (ICCV), 22-29 October 2017, Venice, Italy. Piscataway, NJ, USA: IEEE; 2017. p. 1491–500.

[31] Shokri R, Theodorakopoulos G, Troncoso C. Privacy games along location traces: a game-theoretic framework for optimizing location privacy. ACM Transactions on Privacy and Security (TOPS) 2017;19(4):11.

[32] Watson J. Strategy: an introduction to game theory, vol. 139. New York: WW Norton; 2002.

[33] Wan Z, Vorobeychik Y, Xia W, Clayton EW, Kantarcioglu M, Malin B. Expanding access to large-scale genomic data while promoting privacy: a game theoretic approach. The American Journal of Human Genetics 2017;100(2):316–22.

[34] Basar T, Olsder J. Dynamic noncooperative game theory. 2nd ed., vol. 23. Siam; 1999. Society for Industrial and Applied Mathematics.

[35] Auton A, Abecasis GR, Altshuler DM, Durbin RM, Bentley DR, Chakravarti A, Clark AG, et al. A global reference for human genetic variation. Nature 2015;526: 68–74.

[36] Gaziano JM, Concato J, Brophy M, Fiore L, Pyarajan S, Breeling J, Whitbourne S, et al. Million veteran program: a mega-biobank to study genetic influences on health and disease. Journal of Clinical Epidemiology 2016;70:214–23.

[37] Bowton E, Field JR, Wang S, Schildcrout JS, Van Driest SL, Delaney JT, Cowan J, et al. Biobanks and electronic medical records: enabling cost-effective research. Science Translational Medicine 2014;6(234). 234cm3.

[38] Shringarpure SS, Bustamante CD. Privacy risks from genomic data-sharing beacons. The American Journal of Human Genetics 2015;97(5):631–46.

[39] Fiume M, Cupak M, Keenan S, Rambla J, de la Torre S, Dyke SOM, Brookes AJ, et al. Federated discovery and sharing of genomic data using Beacons. Nature Biotechnology 2019;37(3):220.

[40] Wang S, Jiang X, Tang H, Wang X, Bu D, Carey K, Dyke SOM, Fox D, Jiang C, Lauter K, Malin B, Sofia H, Telenti A, Wang L, Wang W, Ohno-Machado L. A community effort to protect genomic data sharing, collaboration and outsourcing. NPJ Genomic Medicine 2017;2(1):33.

[41] Wan Z, Vorobeychik Y, Kantarcioglu M, Malin B. Controlling the signal: practical privacy protection of genomic data sharing through beacon services. BMC Medical Genomics 2017;10(2):39.
[42] Raisaro JL, Tramer F, Ji Z, Bu D, Zhao Y, Carey K, Lloyd D, Sofia H, Baker D, Flicek P, et al. Addressing beacon Re-identification attacks: quantification and mitigation of privacy risks. Journal of the American Medical Informatics Association 2017;24(4): 799–805.
[43] Hayden EC. Geneticists push for global data-sharing. Nature 2013;498(7452):16.
[44] Craig DW, Goor RM, Wang Z, Paschall J, Ostell J, Feolo M, Sherry ST, Manolio TA. Assessing and managing risk when sharing aggregate genetic variant data. Nature Reviews Genetics 2011;12(10):730.
[45] Humbert M, Ayday E, Hubaux J-P, Telenti A. Addressing the concerns of the lacks family: quantification of kin genomic privacy. In: Proceedings of the 2013 ACM SIGSAC conference on computer & communications security, 4–8 November 2013, Berlin, Germany. New York, NY, USA: ACM; 2013. p. 1141–52.
[46] Humbert M, Ayday E, Hubaux J-P, Telenti A. On non-cooperative genomic privacy. In: International conference on financial cryptography and data security, 26–30 January 2015, San Juan, Puerto Rico. Berlin/Heidelberg, Germany: Springer; 2015. p. 407–26.
[47] Greshake B, Bayer PE, Rausch H, Reda J. OpenSNP–a crowdsourced web resource for personal genomics. PLoS One 2014;9(3):e89204.
[48] Church GM. The personal genome project. Molecular Systems Biology 2005;1(1):1–3.
[49] Wan Z, Vorobeychik Y, Xia W, Clayton EW, Kantarcioglu M, Ganta R, Heatherly R, Malin BA. A game theoretic framework for analyzing re-identification risk. PLoS One 2015;10(3):e0120592.

CHAPTER

Trusted execution environment with Intel SGX

8

Somnath Chakrabarti, Thomas Knauth, Dmitrii Kuvaiskii, Michael Steiner, Mona Vij
Intel Labs, Intel Corporation, Hillsboro, OR, United States

1. Introduction

Big data analysis has become a common trend in many fields. Analyzing aggregated data across organizations has a huge potential benefit to business and research. For instance, if data from hospitals and research institutes were combined, data analytics could have a far greater chance of leading to accurate diagnostics and discovering well-targeted treatments that could improve the quality of life or even save lives. However, the privacy and security of aggregated data are critical due to regulatory requirements and the concerns of mutually distrustful data owners of losing control over these assets. As a result, huge amounts of data remain locked in silos as data owners cannot afford to reveal their data to each other for multiparty analytics.

To break big data-silos and realize the benefit of joint-data analysis, efficient and practical solutions are needed to support privacy-preserving, multiparty analysis, such that:

- Data owners retain control of their data during transfer, storage, and processing: no private data from any data owners shall be leaked; only authorized processing shall be performed and results released only to approved parties.
- There is support to scale and distribute pipelines of dependent jobs in a distributed computation and storage environment to handle the large data size and complexity of real-world analytic workflows.
- The utility of results is not unduly weakened throughout the processing.
- The existing programming and deployment model are supported to ease developing and deploying multiparty analytics applications.

We can decompose the challenge into two orthogonal subproblems:

- secure multiparty computation (MPC), and
- privacy-preserving computation.

The former ensures that the analysis on pooled datasets is executed correctly, and nothing but the final results are revealed and only to authorized parties. The latter ensures that the revealed final results of the joint analysis do not compromise anyone's privacy and helps in satisfying regulatory requirements.

Responsible Genomic Data Sharing. https://doi.org/10.1016/B978-0-12-816197-5.00008-5

The primary challenge with privacy-preserving computation is to balance privacy with utility, the third requirement above. Privacy depends on the type of input data and the algorithmic properties of computation. A common technique to achieve privacy-preserving computation is differential privacy. In its common usage scenario, a trusted server guards the data and executes differentially private algorithms that ensure that (untrusted) clients get only sanitized replies to queries; in particular, no combination of query results will enable the client to deduce any information of individual records of data owners. Although distributed versions of differentially private algorithms exist, they are considerably more challenging to design and deploy than "classical" client/server algorithms. We believe a more promising approach is to convert a client/server differentially private algorithm into a distributed version by running it on top of an MPC environment. In the rest of this chapter, we will focus on how to provide an MPC environment in a scalable and user-friendly manner.

MPC allows a set of N mutually distrustful parties, each with a private input, to securely and jointly evaluate an arbitrary function over their N inputs and to be provided with N outputs, one per party (see Fig. 8.1). No party learns any information from the interaction other than their own output. This is formalized by an "ideal world" abstraction, where there exists an incorruptible trusted party to whom each participant sends its input privately. This trusted party computes the function on its own and sends back privately the appropriate output. Traditional MPC solutions rely on cryptographic protocols to realize a "real world" that provably emulates, based on some cryptographic assumptions, the security properties provided by the "ideal world." In this chapter, such MPC protocols are called as crypto-MPC. Chapter 6 discusses such approaches in detail. However, even with the tremendous progress in practical general-purpose crypto-MPC protocols, they cannot handle the scale of datasets available today, for example, genome data can be measured in terra-bytes and might require days to process even if unprotected. Furthermore, crypto-MPC toolkits require rewriting applications in unfamiliar programming environments and with still fairly immature tooling. This prevents leveraging large existing efforts in big-data analytic tools across different domains, such as the genome analysis toolkit (GATK) [1] widely used in genomics.

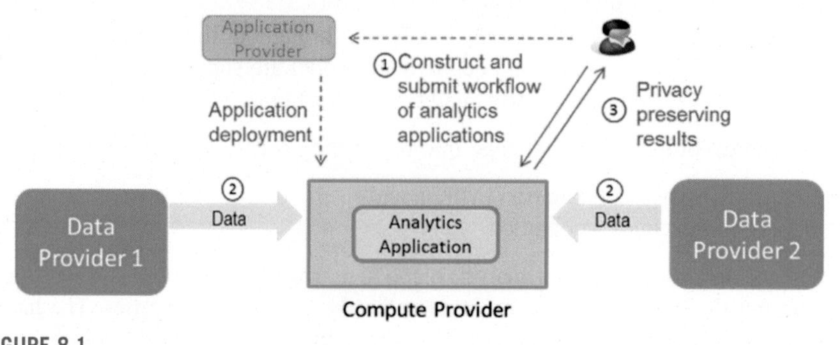

FIGURE 8.1

Multiparty analytics.

To address the earlier issues, we take notice of the rise of trusted computing technologies. In such an approach, trusted execution environments (TEEs) are constructed to host execution such that code outside of the TEE and its trusted computing base (TCB) can neither compromise the execution integrity of the TEE nor the confidentiality of the data processed inside the TEE. Over the years, some hardware security features have been introduced by the industry, such as Intel Software Guard Extension (SGX)/Trusted Execution Technology (TXT) technologies and ARM TrustZone to support setting up TEEs on compute platforms. By using TEEs with hardware-rooted trust, even the cloud provider is moved out of the trust domain. Although no computing system can be completely secure, by leveraging such hardware-based TEEs, one can emulate and provide security properties approaching the "ideal-world" similar to crypto-MPC. We call such a TEE-based approach HW-MPC.

To achieve HW-MPC for a simple system is quite intuitive as a TEE somewhat matches the "ideal world" abstraction of cryptographic MPC protocols. Building scalable MPC solutions on top of TEEs remains a challenge, though. The existing TEE abstraction and interfaces mainly support constructing and attesting TEEs on individual server platforms. Abstractions and services are only emerging on how to (1) compose TEEs to serve complex analytic workflows and meet scalability demand of computation, (2) define and enforce policies for controlling the flow of information (private data) through a workflow of analytic jobs and across TEEs, and (3) provide publicly verifiable proofs for the trustworthiness of the composite TEEs. Developing and deploying TEE applications are nontrivial tasks.

In the following, we will define in more detail the abstractions and properties offered by TEEs (Section 2), explain the realization of the TEE abstraction in Intel SGX (Section 3), explore the deployment of SGX in the cloud (Section 4), and finish with an outlook (Section 5).

2. Trusted execution environment

Execution environments refer to a collection of processors, memory, storage, and peripherals. A TEE provides computation isolated and integrity protected from the normal execution environment. The protections offered by TEEs must be rooted in hardware as it provides minimal TCB and needs to provide protection against hardware as well as software attacks. TCB is the set of hardware and software components that are critical to a TEE's security and must be trusted. The ultimate goal is to minimize the TCB in a TEE, as larger codebases are error prone and difficult to reason about. Fig. 8.2 contrasts the classical untrusted execution environment with a smaller TEE on the example of Intel SGX.

A TEE typically consists of several components: *secure bootstrapping* to help ensure the system starts at a secure initial state; *isolated execution* to protect data and computation; and secure I/O, for example, *sealed storage* to protect data at rest and, in particular, *remote attestation* to allow a remote party to identify and verify

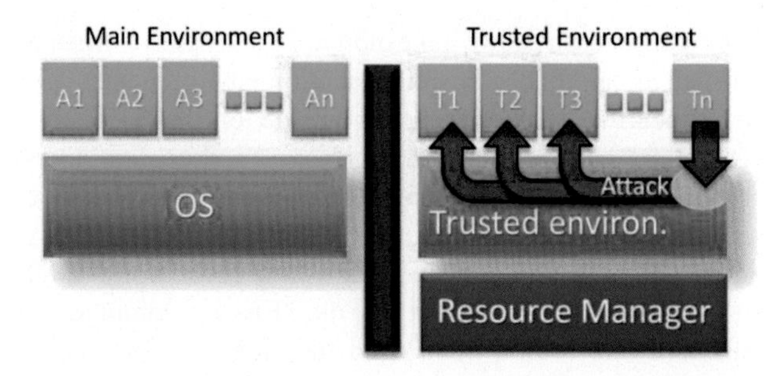

FIGURE 8.2

Classical untrusted execution environment versus trusted execution environment (Intel SGX).

the trustworthiness of a communication peer inside a TEE. Remote attestation enables the establishment of a secure channel that can be used to provision secrets, receive authenticated results, etc. TEEs should provide protections from operators as well as other privileged users accessing restricted data. There is also a strong desire to provide protection against various side-channel attacks and from physical tampering.

TEEs provide a place to stand for building trustworthy software. Some examples of TEEs in production today are Crypto-coprocessors, TEE based on trusted platform module (TPM) and related technologies (Intel TXT, AMD SVM, Intel TME/MKTME, AMD SME), Keystone Enclave for RISC-V, ARM TrustZone, AMD SEV, and Intel SGX.

The earliest secure system in the industry dates back to coprocessors such as IBM 3848 and IBM 4758 [2]. Their purpose was mostly to provide cryptographic processing capability and a means to securely store cryptographic keys. However, the 4758 and later models would also have allowed running arbitrary programs.

Since 2000s, a security coprocessor called *TPM* [3] was standardized by a trusted computing group and widely deployed in PCs. The TPM chip allows the CPU to take and store security measurements of the platform state thereby enabling secure or authenticated boot and subsequent platform integrity of a such-booted TEE. Based on these measurements, TPM's chip-unique RSA key can then be used for platform device authentication and remote attestation. Additionally, TPM provides monotonic counters for rollback security and various other cryptographic services. TPM chips usually also include some security mechanisms to make physical tampering difficult. Originally, TPM required, in a static-root-of-trust, that the BIOS and boot software are part of the TCB. Intel later added *TXT* [4] to CPUs that provide a dynamic root of trust and allows removing BIOS and initial boot software from the TCB. As part of AMD's virtualization extension SVM, an extension similar to TXT was added to AMD chips. Subsequent hardware extensions like *Intel Total Memory Encryption (TME)* and *AMD Secure Memory Encryption (SME)*, allow processor memory to be visible outside of the CPU only in encrypted form and enable further hardening of TPM-based TEEs. They provide *only* heightened confidentiality of data and *not* protection against tampering and rollback attacks.

Furthermore, as the design goal of TPM and related technologies was focused primarily on software adversaries and the overall security against physical adversaries is not very high, TPM-based TEEs might not be sufficient for the need of multiparty security settings where the owner of the computer cannot necessarily be trusted. Additionally, TPM-based TEEs always include an operating system or hypervisor in the TCB and hence have fairly large TCBs. This makes it challenging to get an assurance of code correctness high enough for sensitive operations such as processing of medical data.

The Keystone project [5] is an open-source TEE implementation for RISC-V processors. It was unveiled in 2018 and aims to provide an open-source, highly customizable TEE that can be optimized for resource usage of a particular application. Keystone is influenced by the previous commercial designs including ARM TrustZone and Intel SGX. Therefore, it implements the same TEE components as other techniques, including remote attestation, memory isolation, secure bootstrapping, and a security monitor. Different from Intel SGX, Keystone does not currently provide a memory encryption/integrity engine though such a component can be implemented as an extension. As of this writing, the Keystone project is in its early stages of development (version 0.1), and its performance, usability, deployability, and other characteristics are still unclear.

ARM TrustZone [6] goes beyond coprocessors and enables the development of isolated execution environments directly in the main CPU by supporting two domains (known as worlds): Normal and Secure. Normal operating systems and applications run in the normal domain; trusted OS and applications run in a secure domain. These domains are isolated by the CPU and components running in a normal domain cannot access any memory or other resources running in the secure domain. This model provides strong isolation enforced by hardware but does not scale well as there is only a single TEE. Running inside the secure domain a secure OS, which provides, for example, GlobalPlatform [7] based TEEs, can address this but does so at the cost of increased TCB size and, depending on OS and TEE abstractions, a restricted programming environment. This together with the fact that TrustZone is more widely available only on mobile devices but not on server systems makes TrustZone less suitable for big data analytics in multiparty security settings.

AMD Secure Encrypted Virtualization (SEV) [8] extends SME to AMD virtualization, allowing individual VMs to run SME using their own secure keys and providing remote attestation. Similar to TPM-based TEEs, it requires large TCBs including a complete virtual machine monitor (VMM). Although SEV provides confidentiality, it does not provide strong integrity protection, in particular, no protection against replay attacks. Additionally, applying SEV in the context of collaborative analytics requires overcoming the challenge that SEV's remote attestation is a-priori verifiable only by the provider of the VM and is not publicly verifiable, as required in multiparty security settings.

Intel SGX [9] provides a trusted execution environment that is intended to provide a scalable and attestable secure execution environment in a mainstream platform. In the following sections, we will describe Intel SGX in detail, explore the deployment of SGX in the cloud, and finish with an outlook.

3. Intel Software Guard Extensions

In 2015, Intel introduced a new secure extension to the upcoming Skylake CPUs called Intel SGX. Intel SGX is an actual hardware realization of a TEE providing confidentiality and integrity protection, with a particular focus on small TCB, support for legacy applications, ease of application development and deployment, and acceptable performance overhead.

Intel SGX was designed to protect secure computations even in the presence of a powerful attacker. The attacker can control the whole software stack including the operating system and the hypervisor. She can also have access to the physical machine: she can observe and modify all data in main memory (RAM), snoop on the memory bus, and control I/O devices. The *only* hardware part that the attacker is unlikely to hijack is the CPU package. Thus, with Intel SGX, the end user needs to trust only the Intel CPU.

The central concept in Intel SGX is the so-called *enclave*. Enclaves are opaque regions of memory carved out of the normal application (process). They securely execute sensitive code on sensitive data, such that no other part of the "host" process, nor any other process, nor even the privileged software (operating system, hypervisor) can access enclave data or subvert enclave code execution (see Fig. 8.3).

Importantly, SGX enclaves execute on the same hardware as other software. An SGX enclave can be thought of as a "privileged" part of a user application: the application performs its normal noncritical computations most of the time, but for sensitive critical computations it switches to the enclave mode (enters the enclave), the enclave securely runs these computations and then exits, thus switching back to the normal application mode.

From the hardware perspective, Intel SGX introduces several new features. First, a new CPU mode: whenever a user process wants to enter an enclave, the CPU

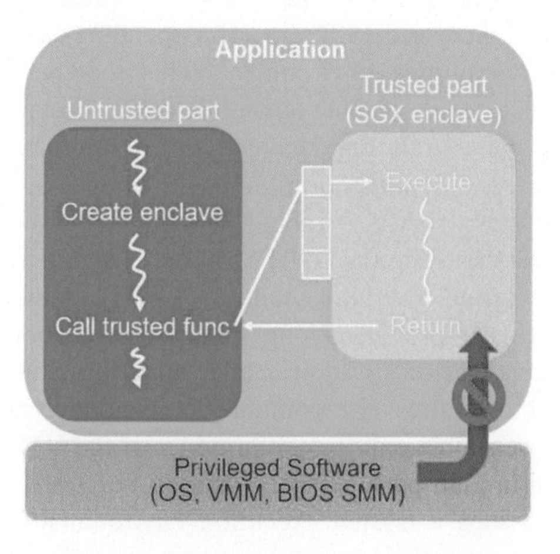

FIGURE 8.3

Interaction between untrusted and trusted parts of the application in Intel SGX.

switches to the enclave mode. In this mode, only the enclave code is allowed to execute, some security-sensitive CPU instructions are forbidden, and all enclave data are transparently encrypted/decrypted when moved out/to the CPU caches. The second new feature is an enclave page cache (EPC)—a region of RAM dedicated to storing enclave code and data pages. The enhanced memory access controller on the CPU allows memory accesses inside the EPC only for the corresponding SGX enclave, preventing attacks from other applications and privileged software. EPC also contains enclave-specific metadata to prevent subtle attacks discussed in the next sections. The third new feature of Intel SGX is the memory encryption engine (MEE)—a separate component on a CPU that transparently encrypts all data traveling from CPU to EPC and transparently decrypts data in another direction. MEE uses cryptographic keys for encryption to help provide confidentiality, message authentication codes to provide integrity, and versioning to provide freshness, for example, protect against rollback attacks.

From the software perspective, Intel SGX provides a familiar development and deployment environment. Recall that pure software solutions for secure computations (homomorphic encryption, multiparty computations) rely on complex cryptographic protocols that incur high-performance overheads and are built using specific, limited software primitives. In contrast, SGX enclaves execute normal $\times 86$ instructions on a CPU and compute over plaintext data in CPU caches. This allows to run legacy unmodified code inside enclaves over normal unencrypted data with near-native performance. (As some instructions are forbidden inside SGX enclaves, it may be sometimes necessary to adapt legacy code; see next sections.) In fact, enclaves are usually programmed as *shared libraries* and loaded inside the EPC at process initialization so that the rest of the application invokes library functions to execute secure in-enclave functionality.

Finally, the end user (who may want to run secure computations on a remote machine in another part of the globe) must gain trust in enclave execution. In particular, the user wants to ensure that the enclave executes the code she expects/trusts and that the enclave executes on a real SGX-enabled Intel CPU. To this end, Intel SGX provides a cryptographic measurement over all code and data at the enclave initialization as well as a special "report" on the CPU signed by a CPU-specific key. With the help of the remote attestation (RA) procedure of Intel SGX, the user can obtain these measurements, compare and verify them against the expected values, and by doing so gain trust in the remote enclave.

To put it all together, we will use the example of an aggregate association analysis application over shared genomic data. In our scenario, illustrated in Fig. 8.4, several parties owning genomic data are pooling their datasets to allow for higher-quality aggregate association analysis. However, as the data providers do not completely trust each other nor the researchers performing the analysis, the computation is performed in a multiparty secure manner inside an SGX enclave. In preparation (step 0), the application is provisioned and the (large) data-sets are encrypted and uploaded. To perform an analysis, the analyst launches the analysis application in an enclave at the cloud provider (step 1). After assuring themselves

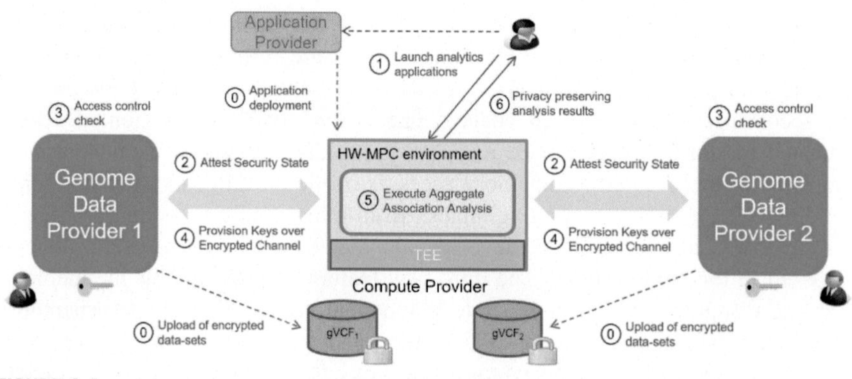

FIGURE 8.4

Example application performing collaboratively aggregate association analysis over genomic data.

that the application is genuine and appropriately protects their privacy as well as approving all involved parties (steps 2 and 3), the data providers release the decryption keys for the previously uploaded encrypted data files containing the DNA sequences to the enclave (step 4). Having now access to the data, the enclave performs the analysis (step 5) and returns the results to the analyst (step 6). Thanks to transparent memory encryption in hardware, all genomic data and all intermediate computations are confidentiality- and integrity-protected. With the help of remote attestation, the data providers and analysts can gain trust in the application code and the Intel SGX platform. This allows the establishment of a secure channel to the enclave so the decryption keys will be visible only to the enclave and the result can be authenticated and privately delivered to the analyst, ensuring the required multiparty security requirements.

3.1 Hardware architecture

Intel SGX provides a secure execution environment for applications using a standard programming model. However, Intel SGX does not trust any hardware or software components for its operation except the CPU package but strives to provide a secure and trusted execution environment. To achieve this level of protection, several new capabilities have been added to the Intel CPU architecture.

As previously mentioned, when enclave code and data reside inside registers, caches, or other logic blocks within the processor package, then any unauthorized access may be prevented. This is done by access control mechanisms built inside the CPU's SGX microcode. In addition, when enclave data leave the CPU caches to be written into protected main memory (EPC), the data are automatically encrypted and integrity protected preventing memory probes or other techniques to view, modify, or replay data or code contained within an enclave. This is done by the memory encryption engine. Fig. 8.5 shows a high-level hardware architecture of Intel SGX, with the CPU package as the trust boundary.

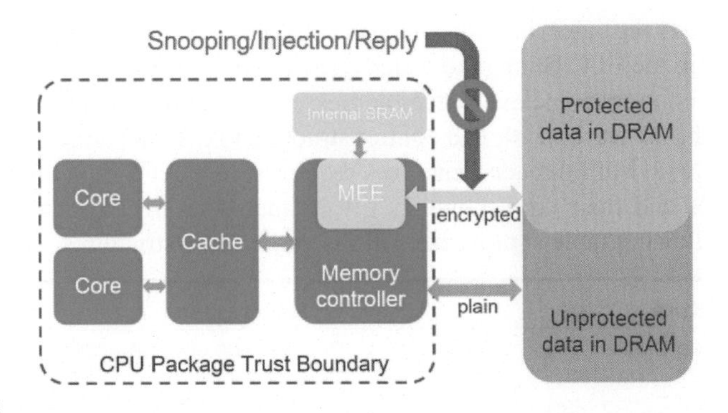

FIGURE 8.5

Intel SGX: hardware architecture.

3.1.1 Memory encryption engine and protected memory region

The MEE is a hardware unit on the CPU package that encrypts and integrity protects enclave traffic between the processor package and the main memory (DRAM). The overall memory region that an MEE protects is called the MEE region and is protected using CPU PRMRR (processor reserved memory range register). The MEE uses a single encryption key to help protect all enclaves on the platform that is automatically generated every time the CPU resets. Thus, attempts to modify an enclave's contents are detected and either prevented (during software attacks) or execution is aborted (during hardware attacks).

An SGX enabled Intel CPU implements the MEE region as a cryptographically protected DRAM region and supports the ability for the BIOS to reserve a range of memory called processor reserved memory (PRM). The BIOS allocates the PRM by configuring a set of range registers, collectively known as the PRMRR.

3.1.2 Enclave page cache and SGX data structures

To implement SGX memory protections, new hardware logic, microcode, and data structures are required. In particular, Intel SGX introduces the EPC. It is a protected memory region where enclave pages and SGX data structures are stored. This memory is designed to be protected from unauthorized hardware and software access. Code and data from all the enclaves on the platform reside inside the EPC. When an enclave performs a memory access to the EPC, the processor decides whether or not to allow the access. The processor maintains security and access control information for every page in the EPC in a hardware structure called the enclave page cache map (EPCM). This structure is consulted by the CPU and SGX instructions only, and it is not accessible by code running inside enclaves. The security attributes for each EPC page are held in this internal microarchitecture data structure.

The EPC is divided into 4KB chunks called EPC pages. EPC pages can either be valid or invalid, and a valid EPC page contains either an enclave page or an SGX structure.

Each enclave instance has an enclave control structure, SECS. Every valid enclave page in the EPC belongs to exactly one enclave instance. The system software is required to map enclave virtual addresses to a valid EPC page. In addition, each enclave has at least one thread control structure (TCS) that stores metadata for an enclave thread. Multithreaded enclaves require several TCS structures.

EPC pages and their corresponding EPCM metadata are physically stored in PRM. The following table summarizes all the earlier data structures.

Data structure	Description
Enclave Page Cache (EPC)	Contains enclave application code and data
Enclave Page Cache Map (EPCM)	Contains metadata of all enclave pages
SGX Enclave Control Structure (SECS)	Metadata for each enclave
Thread Control Structure (TCS)	Metadata for each thread

3.1.3 Enclave creation and execution

The enclave creation starts by compiling the enclave code into a binary library. The binary is then loaded into the EPC and is assigned a unique enclave identity on the platform.

The enclave creation process is divided into multiple stages: initialization of enclave control structure, allocation of EPC pages and loading of contents into the pages, measurement of the enclave contents, and finally establishing the enclave identity.

These steps are supported by the following new CPU instructions: ECREATE, EADD, EEXTEND, and EINIT. Additionally, EENTER, EEXIT, and ERESUME are used to enter and exit the enclave (described in the following). The new CPU instructions are summarized in the following table.

Instruction	Description
ECREATE	Create SECS page
EADD	Add enclave pages
EEXTEND	Measure 256 bytes of enclave page
EINIT	Finalize enclave
EENTER	Enter into enclave at predefined entry point
EEXIT	Exit enclave
ERESUME	Enter into enclave after serving exception/interrupt

ECREATE starts the enclave creation process and initializes the SGX enclave control structure (SECS) that contains global information about the enclave. EADD commits EPC pages to an enclave and records the commitment (but not the contents) in the SECS. The memory contents of an enclave are explicitly measured by EEXTEND.

Once the process of building the enclave in EPC is completed, the enclave needs to be securely locked down. Locking down an enclave in the EPC helps prevent any further changes or additions to enclave pages from outside the enclave. The CPU will only allow entry to the enclave once it has been locked down. The finalization process is executed by the EINIT instruction. Upon successful completion of the EINIT instruction, the enclave's cryptographic signature is recorded by the CPU. This signature uniquely identifies the code and initial data loaded inside the enclave.

Intel SGX also introduces EENTER and EEXIT instructions, to enter and exit an enclave programmatically. Enclave entry and execution can start only at fixed predefined entry points. Enclave exits, however, can occur due to a fault or an interrupt. Upon such asynchronous exits, the processor invokes a special internal routine called asynchronous exit (AEX) that saves the enclave state inside the enclave, loads a synthetic state in CPU registers, and sets the faulting instruction address to a value specified during EENTER. After the fault/interrupt has been serviced by the OS, it executes the ERESUME instruction that restores the state back to allow the enclave to resume execution.

3.1.4 Enclave paging

Intel SGX architecture also offers instructions to allow system software to oversubscribe the EPC by securely evicting and loading enclave pages and SGX data structures. This enables the system software to be able to run a large number of enclaves on a platform even if the total amount of enclave memory required is larger than the EPC size available on the platform. To achieve this, system software first identifies less frequently accessed EPC pages and evicts them to make space for more active EPC pages.

Intel SGX architecture enables the contents of an enclave page evicted from the EPC to main memory to have the same level of integrity, confidentiality, and replay protection as the contents residing inside the EPC.

To achieve the same security level, SGX architecture ensures that the following conditions are met while evicting or loading EPC pages:

- Enclave pages are evicted only after all CPU cache translations to that page have been cleared from all CPU cores.
- The contents of all evicted enclave pages are encrypted and integrity protected before being written out to main memory.
- When an evicted enclave page is reloaded into EPC it must have the same page type, permissions, virtual address, and content, and belong to the same enclave as at the time of eviction.
- Only the latest evicted version of an enclave page is allowed to be reloaded. Old copies of the evicted pages are invalidated every time a new eviction of a page is made.

The following table shows all SGX instructions used for paging and their use.

Paging instruction	Description
EPA	Create version page to track the latest version of evicted pages
ELD	Load evicted page from regular memory into EPC
EWB	Evict EPC page to regular memory
EBLOCK	Block EPC page from further access
ETRACK	Track all old CPU address translations to the EPC page
EREMOVE	Mark the EPC page as INVALID in EPCM

3.1.5 Enclave teardown

Once the enclave execution is over and the application no longer needs the enclave, it can be permanently removed from the EPC. Similar to the creation process, the removal process also must follow specific steps to maintain SGX security properties. The first step involves system software ensuring no threads are currently executing inside the enclave followed by removal of all process page table mappings to the enclave EPC pages. This will ensure that no further execution can take place inside the enclave. At this stage, all regular enclave pages (except SECS page) can be removed by system software using EREMOVE instruction.

EREMOVE instruction will mark the page as an INVALID page in EPCM and adjust the child page count number in the parent SECS page. Note that once an enclave page is removed using EREMOVE instruction it cannot be added again to the enclave and the only way the page can be added again is by recreating the enclave. The SECS page must be the last page to be removed from the EPC as EREMOVE instruction ensures that all child enclave pages are removed before removing the SECS page.

3.1.6 SGX platform stack

Fig. 8.6 shows a high-level hardware/software architecture of the SGX stack on a platform. Enclaves are the trusted parts of applications, communicating with the outside world via SGX user runtime. Page tables for enclaves are managed by the untrusted OS. In the hardware, enclave pages are mapped to EPC, and the EPCM is used to enforce the access control to EPC memory. All SGX instructions that are used to manage enclaves are actually leaves of two real CPU instructions, namely ENCLU and ENCLS. ENCLU is the user-level instruction that allows applications to enter/exit/resume the enclave and manages enclave keys. ENCLS instruction can only be executed by privileged software, which is used to create/remove/measure the enclave as well as perform paging operations.

3.1.7 Remote attestation

Another important aspect of Intel SGX is the support for remote attestation. Remote attestation provides the ability to remotely gain trust in an enclave and enables the remote party to share sensitive code, data, and secrets with trust and confidence using an authenticated and secure channel.

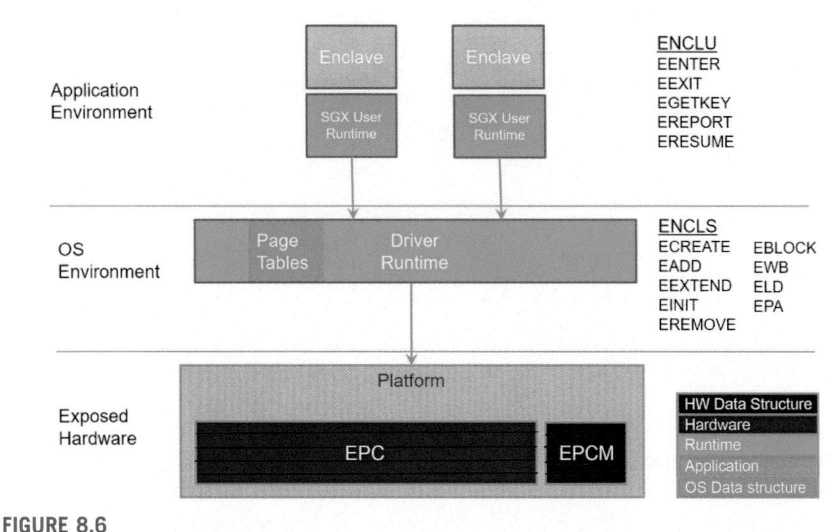

FIGURE 8.6

SGX platform stack and new instructions.

The process of establishing an authenticated and secure channel starts by generating a hardware-based attestation report identifying the enclave software as well as other security properties of the enclave including CPU's hardware and firmware properties. The measurement of the software, that is, code and data inside the enclave, is a vital parameter in the attestation report and is basically a hash digest over enclave memory as well as enclave configuration. Any attempt to change the code, data or enclave configuration will result in a different hash digest reflected in the attestation report.

Once generated, this hardware-based attestation report is then verified and signed as a quote by an Intel developed enclave called Quoting Enclave running on the same platform. This quote can then be cryptographically verified by the user at the remote end to gather authentic information about the software and the environment the enclave is running on.

The general flow of remote attestation is depicted in Fig. 8.7. An example of remote attestation and secure channel establishment with an SGX enclave is as follows:

- Once the application enclave is loaded on a platform, the remote user establishes a communication channel with the application and challenges the application enclave to prove that it is indeed the trusted enclave. Note that this initial communication channel is considered untrusted until a successful remote attestation takes place.
- The application enclave creates a manifest containing the response to the challenge and a self-generated ephemeral public key to be used by the remote user for secure communication.
- The enclave executes the EREPORT instruction with the hash digest of the manifest as an input to the instruction. The instruction generates the REPORT that binds the manifest to the enclave. The enclave sends the REPORT and the

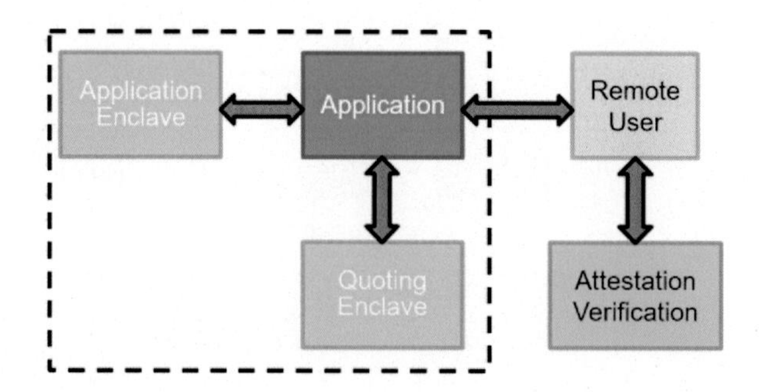

FIGURE 8.7

Remote attestation flow.

manifest to the application and it forwards the REPORT to the Quoting Enclave for signing.

- The Quoting Enclave (see Fig. 8.8) retrieves its Report Key using the EGETKEY instruction and verifies the REPORT. The Quoting Enclave then creates the QUOTE structure—which among other things contains the hash of enclave manifest, the enclave's identity and information on the platform TCB and QE identity—and signs it with its EPID key. The Quoting Enclave returns the signed QUOTE back to the application.
- The application sends the signed QUOTE and the manifest to the remote user.

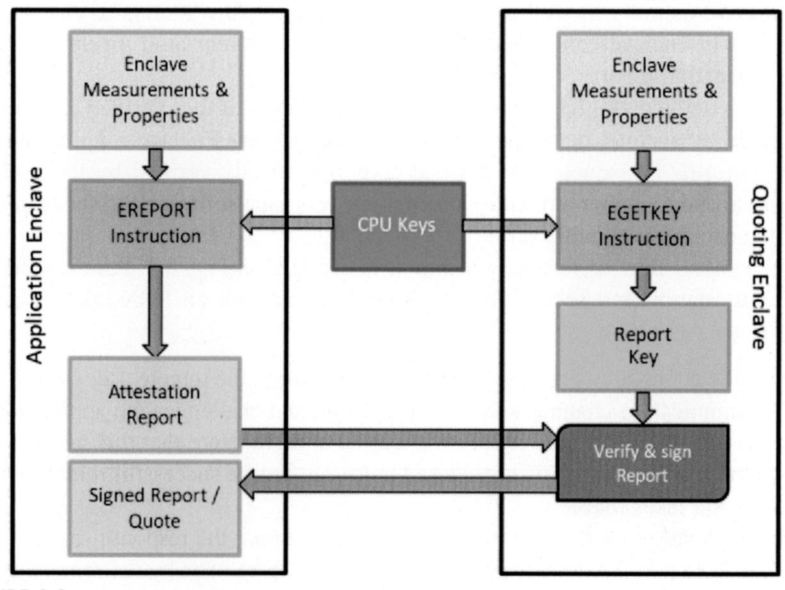

FIGURE 8.8

Signing of enclave's REPORT by Quoting Enclave.

- The remote user, with the assistance of an attestation verification service (Intel IAS by default), validates the signature of the quote. This attestation verification service is the ultimate root of trust in the enclave and the platform it runs on and provides, for example, information of up-to-date platform TCB versions and of revoked platforms. Based on this information, one can determine whether the particular platform and QE can be trusted and, if so, conclude that the (hash of the) manifest must come from an enclave with identity and properties as specified by the signature content.
- The remote user verifies the integrity of the manifest using the embedded hash digest and uses the data in the manifest as the response to the challenge that was sent in the first step.
- Once the remote user/entity is satisfied by the attested identity of the application enclave, it continues with the secure key exchange to complete the secure channel setup.

The following table describes two SGX instructions used for remote attestation.

Attestation instructions	Description
EREPORT	Generate REPORT containing enclave parameters like measurement, enclave signing key hash, and enclave properties
EGETKEY	Derive enclave- and platform-specific keys for various purposes, including the key for attestation (Report Key)

Remote attestation is an integral part of the deployment of our aggregate association analysis application (see Fig. 8.4). In particular, all the involved parties must attest the enclave: all data providers and the analyst must gain trust in the aggregate association analysis application code and the Intel SGX platform it runs on. For each involved party, remote attestation proceeds as described earlier. At the end of this process, each data provider as well as the analyst establish a secure channel to the enclave and provide it with data to perform the desired multiparty computation. For more information on remote attestation and the different types of remote attestation supported, see Refs. [10,11]. See Ref. [12] for further information on how to integrate it into secure channels.

3.2 Software development

The programming model for SGX enclaves is similar to the usual shared-library development. In particular, all sensitive code and data must be isolated from the rest of the application in a self-contained library with a few well-defined entry points. At runtime, this self-contained library is measured, loaded inside an SGX enclave, and remotely attested by end users of the applications. The application in which the enclave is embedded is called as the *host application* and is considered untrusted (see Fig. 8.9).

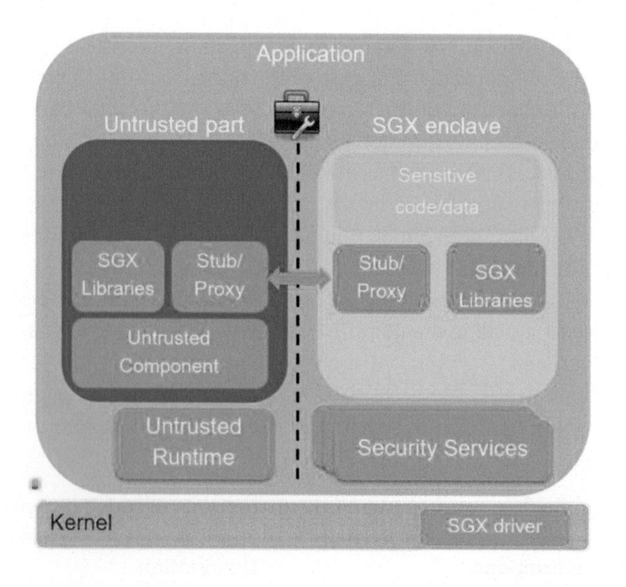

FIGURE 8.9

Intel SGX: software architecture.

The usual runtime flow is as follows. When the host application is started, it creates the enclave, populates it with sensitive code and initial data, measures the enclave, and starts it. After this initialization phase, the enclave is ready for secure computations. At runtime, whenever the host application needs to run enclave code, it passes control to the enclave by entering it, the enclave executes and exits back to the host application. At shutdown, the host application is responsible for the graceful enclave teardown.

As enclaves are executed on Intel CPU, legacy code can be ported to enclave code. Unfortunately, some CPU features are unsafe in enclave mode and are thus disabled, making it impossible to invoke system calls (to instruct the OS to send a network packet, for example) from inside the enclave. Therefore, if an enclave needs some OS functionality, it must pause its computations, exit so that the untrusted part of the application performs a system call on its behalf, and be reentered to continue execution with the result of system call. All enclave communication must be channeled through the address space of the host application.

Just like shared libraries, enclaves must be entered only through a well-defined API. The entry points (also called *entry functions* or *call gates*) of the enclave are fixed by SGX, such that the malicious host application cannot simply jump to an arbitrary point in enclave code and disrupt normal execution flow. The act of entering the enclave through one of the entry points is called *ECALL*, or enclave call. Similarly, the enclave may exit to ask the host application to perform some untrusted computation—this is called as *OCALL*, or outside call. ECALLs are akin to normal library function calls, while OCALLs are akin to callbacks. Input arguments and output results are communicated between the untrusted host

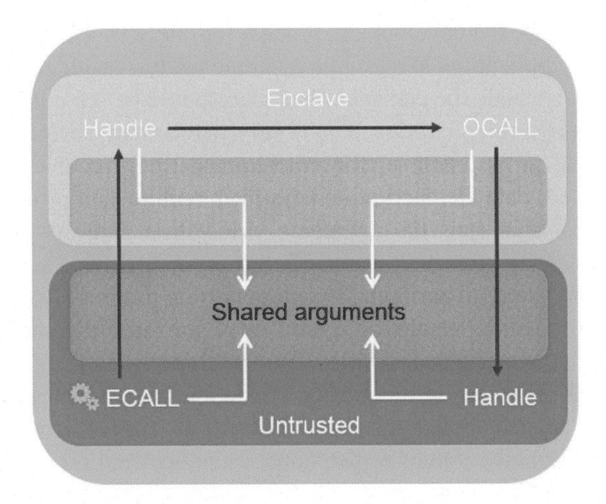

FIGURE 8.10

OCALL and ECALL interfaces in Intel SGX.

application and the enclave through untrusted shared memory. The flow of ECALLs and OCALLs is summarized in Fig. 8.10.

The internal mechanics of ECALLs and OCALLs are similar to Remote Procedure Calls (RPCs) and include complicated marshaling/unmarshalling of arguments, their verification, and enclave entry/exit. For this reason, special software tools simplify the development of the ECALL/OCALL interface, automatically generating stubs/proxies from the user-supplied high-level interface description (this is similar to, e.g., protobuf [13]).

One such tool is the Intel SGX software development kit (SDK) tailored for C/C++ applications. It provides the "edge interface" generator that creates ECALL/OCALL boilerplate code and links it into the host application. It also provides the "trusted runtime" used inside the enclave: helper SGX libraries for trusted memory management, file I/O, encryption mechanisms, random number generator, sealing of data on persistent storage, etc. The SGX SDK also provides the "untrusted runtime" used by the host application: helper SGX libraries to load, start, initialize, and measure enclaves (more specifically, these libraries instruct the SGX kernel driver to perform these actions). Finally, the SGX SDK provides additional security services such as enclave configuration, in-enclave debugger, and IDE extensions, as well as generating the expected measurement of the enclave used by the end user as a reference during remote attestation. In general, Intel SGX SDK allows developers to create and port C/C++ enclave applications with minimal effort.

SGX enclaves can only be started on an SGX-enabled hardware. On hardware without SGX support, any attempt to start the enclave will fail. The attacker can still try to emulate the SGX enclave and SGX hardware, but remote attestation in emulation mode will fail to verify.

Operations to create, initialize, and measure enclaves must be done by privileged software and are provided via an SGX driver. SGX drivers must be installed to use these operations.

Let us now describe the development process for our SGX-enabled aggregate association analysis application. We assume a simple analysis with an existing library that can be put as-is inside the enclave. This code should be separated from the rest of the application in a shared library. The library will contain several ECALL functions, for example, "prove_enclave" for remote attestation, "receive_keys" that provides the (encrypted) data file decryption keys from remote data-providers as inputs, and "run_analysis" that runs the aggregate association analysis and returns the (encrypted and integrity protected) result. The library will also use several OCALLs, for example, "print_stats" to periodically output current progress, "persist_analysis" to checkpoint intermediate results on a hard drive for fault tolerance, and "read_-data" to access the encrypted data files. The developer uses the "edge interface" generator of SGX SDK to automatically create the necessary proxy code. The untrusted part of the host application will only serve as a proxy between the enclave and the user. Additionally, it may output statistics and current progress for the local administrator. In-enclave code must call SGX SDK helpers to perform sealing, generate encryption keys, etc. Because system calls are prohibited inside the enclave, our in-enclave library must not try to directly communicate with the operating system. Finally, the resulting library must be measured by the SGX SDK and can be shipped to remote machines for secure execution.

Many bioinformatics applications are much more complicated than our simple aggregate association analysis example. They may invoke system calls extensively, perform many I/O operations, and request trusted counters/timestamps. In this case, partitioning and porting applications to SGX enclaves may be challenging and time consuming, even with the help of Intel SGX SDK.

Fortunately, there are several existing frameworks to run unmodified legacy applications inside SGX enclaves, including SCONE [14], SGX-LKL [15], and Graphene-SGX [16]. They were shown to achieve good trade-off in terms of security, porting effort, and performance. Using a framework like Graphene-SGX, our aggregate association analysis could be run as-is inside the enclave, with no modifications required. This "push-button enclavization" comes at a price of a considerably increased TCB, possibly larger attack surface, and slight degradation in performance.

3.3 Security properties

Thanks to its minimal hardware TCB of only the CPU package, Intel SGX can thwart an array of hardware and software attacks. The general attack vectors are shown in Fig. 8.11. In the following, we outline the attacks and explain how the SGX architecture prevents them.

Hardware attacks. Intel SGX is one of the few TEE technologies that can mitigate direct hardware attacks. The attacker may have complete control over the physical machine and can attach malicious devices to, for example, snoop on the memory bus. Alternatively, the attacker can launch a DMA attack by attaching her device to an expansion port to directly observe physical memory. The underlying

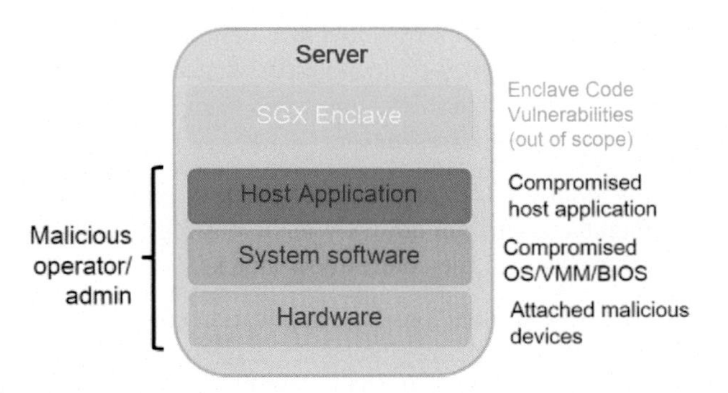

FIGURE 8.11

Attack model of Intel SGX.

memory encryption engine for Intel SGX stores all enclave data encrypted and integrity-protected in memory, thus complicating hardware attempts to read or modify enclave contents.

System software attacks. Intel SGX can protect against a range of attacks by privileged system software: firmware, hypervisors, and operating systems.

- *Reading/modifying physical memory.* Modern computers are shipped with omnipotent firmware, invisible even to operating systems and hypervisors. This firmware (system management mode and BIOS/UEFI) is a frequent target of attacks. Such compromised firmware may read and modify all physical memory of a computer. Similarly, OSes and VMMs (or hypervisors) are privileged software that have full access to both virtual and physical memory of applications. Thus, a compromised OS or VMM may read and overwrite any application data. Intel SGX helps protect against system software's attempts to read or modify enclave data because SGX strives to protect all enclave data leaving the CPU package with encryption. In addition, at the software level, access to EPC contents is allowed only for the corresponding enclave; SGX disallows access to EPC for any other software.
- *Address translation attacks.* Malicious operating systems and VMMs may launch more insidious attacks on the enclave by tampering with page table mappings. Recall that the OS is responsible for the virtual-to-physical-pages mapping of application pages (this mapping is removed when a physical page is evicted from RAM to secondary storage and recreated when the physical page is swapped back into RAM). The malicious OS could silently jumble up these mappings and subvert enclave's execution flow. Intel SGX introduces a special mechanism to verify that the OS maintains correct page tables, therefore preventing this indirect attack.
- *Replay attacks.* As the attacker cannot modify enclave data due to SGX encryption, she can try to feed the enclave its own old data in the hope of subverting the enclave's execution. For example, the attacker might replace intermediary statistics for the aggregate association analysis with previously

computed stale one and the enclave would then compute misleading results. To prevent this attack, the integrity-protection mechanism of SGX generates unique version numbers and verifies them when bringing enclave data back to the CPU (thus achieving data freshness).

Host application attacks. The host application that runs the enclave can be compromised by the attacker to run arbitrary malicious logic. The possible attacks on the enclave, in this case, are the control flow attacks.

- *Control flow attacks*. The malicious host application can try to subvert enclave execution by jumping not to the beginning of the requested enclave entry, but to some arbitrary place in enclave code. This could lead the enclave to "forget" to perform some computations. Intel SGX helps prevent such random jumps, allowing to enter the enclave only at a predefined set of entry points.

Attacks by malicious actors. All the earlier attacks can be launched by an insider operator who has full access to both hardware and software running on the physical machine. An additional attack vector is impersonating the SGX enclave on an attacker-controlled machine and directing end users to it.

- *Malicious operator/administrator*. A malicious operator or administrator has physical access to the machine where the user enclave runs and has complete control over its privileged software. Thus, the operator can launch any of the aforementioned attacks. As we discussed earlier, however, all these attacks are undermined by Intel SGX.
- *Impersonation attack*. A malicious remote party may pretend to run a secure enclave on a genuine Intel SGX-enabled machine, while in reality it would emulate enclave execution and apply all attacks described earlier. However, the remote attestation procedure of Intel SGX ties the identity of the server and the enclave to the unique secret known only to the SGX-enabled CPU. Thus, for a successful impersonation attack, the hacker needs to obtain/guess the secret that is practically impossible.

Although current Intel SGX implementations help prevent a wide range of malicious behaviors, they are still susceptible to the following attacks:

System software attacks. Although Intel SGX protects against most of the attacks launched by privileged system software, some attacks are out of the scope of the HW protection. Enclave developers and users must be aware of these attacks and harden SGX applications as deemed appropriate.

- *Denial-of-service attacks*. As all privileged software is not in the TCB, the server can simply refuse to run the enclave or pause it frequently. This would lead to deteriorated QoS guarantees or complete denial of service. This attack is out of SGX scope, but we note that it is also not in the interest of a service provider.
- *Iago attacks*. Any SGX enclave needs to communicate with the outside world. At the very least, the enclave relies on network I/O from the underlying operating system to receive requests from the remote user and send his replies back.

This is achieved by ECALLs/OCALLs with specific arguments. However, the untrusted host application/OS can tamper with arguments and return values of these calls, which in some cases may subvert enclave execution [17]. Enclave developers must design the ECALL/OCALL interface carefully and double-check all potentially malicious return values to thwart Iago attacks.
- *Side-channel attacks*. Side-channel attacks are attacks that do not access sensitive data directly but can infer it to some degree by observing changes in the microarchitectural state. Powerful side-channel attacks such as cache attacks [18], timing attacks [19], Rowhammer DRAM attacks [20] are especially threatening in the SGX context as they leak enclave-protected data. Intel SGX is also susceptible to a specific type of side-channel attack called controlled-channel attacks: a malicious operating system can induce page faults on every page accessed by the enclave and observe enclave memory accesses at page granularity [21]. Intel CPUs (including SGX-enabled CPUs) provide microcode updates to patch Intel hardware against some of these attacks. Various software techniques help mitigate these attacks.

Attacks on enclaves. Finally, the SGX technology provides no security guarantees regarding code running inside the enclaves themselves. Thus, attackers can still launch attacks on buggy software executing inside an SGX enclave.

- *Vulnerabilities in enclave code*. Intel SGX helps protect the enclave from the malicious outside world, but it cannot protect a malicious enclave from itself. If the enclave code contains bugs that can be exploited by an attacker, SGX provides no confidentiality and integrity guarantees. Enclave developers must rely on well-established techniques for software reliability such as code reviews, formal verification, extensive testing, dynamic memory safety [22], and CFI/CET [23].

3.4 Performance properties

Thanks to the fact that Intel SGX executes native enclave code directly on a CPU using plaintext enclave data in CPU caches, it significantly outperforms software-only cryptographic schemes such as fully homomorphic encryption, garbled circuits, and MPC. For example, the PRINCESS [24] privacy-preserving genomic data analysis framework based on Intel SGX performs orders of magnitude better than homomorphic encryption and garbled circuit solutions.

Improving confidentiality and integrity of applications always comes at a cost of performance degradation. In the case of Intel SGX, there are three main sources of performance overhead.

First, additional hardware checks on memory accesses inside EPC, as well as transparent encryption, authentication, and replay protection of EPC pages inevitably degrade performance. Note that cache-hit accesses incur no performance penalty because they do not propagate outside of the CPU package, while cache-misses are already rather expensive and thus amortize the overhead of EPC accesses.

The second source of overhead is transitioning between the enclave and the host application. Entering and exiting the enclave is notoriously slow, and frequent ECALLs/OCALLs can lead to high overheads. However, this problem can be solved by a switchless design where the enclave never exits and runs in parallel to the host application [25]. In detail, the enclave runs on one CPU core and communicates with the untrusted application running on another CPU core via a shared queue where both store ECALL/OCALL requests and responses. This switchless design can be enabled in most SGX tools including SGX SDK [25], Graphene-SGX [16], SCONE [14], and SGX-LKL [15].

The third source of overhead stems from a limited size of EPC. If the working set of the enclave exceeds the size of the EPC, a slow EPC paging mechanism is employed. This can be a serious performance bottleneck for genomics applications that operate on huge datasets. To alleviate this issue, the developer must employ locality-aware techniques to operate on small chunks of data at a time (similar to cache blocking).

4. HW-MPC and SGX in cloud

Bioinformatics today tackles problems that are infeasible to compute on a single machine. Therefore, bioinformatics applications require cost-effective scaling where thousands of (virtual) machines are employed to collaboratively compute over application data. The *cloud computing* paradigm emerged to solve the problem of agile and cost-efficient scalability. With cloud computing, a large fleet of virtual machines can be acquired in a short time and decommissioned after the problem at hand is solved.

Unfortunately, a user who wants to offload computations to the cloud must trust the cloud service provider. Numerous attacks on cloud platforms raise privacy concerns over offloaded user data [26]. Trusted hardware architectures such as MPC or TEEs bring back control over private computations in untrusted clouds to the end user. Among the trusted execution environments surveyed earlier, Intel SGX is unique in the security properties, versatility, and performance it provides. Due to its high performance and minimal TCB comprising only the CPU chip and the enclave, Intel SGX is a perfect match for untrusted cloud environments. Even a cloud administrator with physical access to a cloud server cannot deduce what code and data are running inside an enclave, whereas this remains a threat with cloud providers that do not offer Intel SGX.

To make Intel SGX in the cloud a viable option, it must integrate with existing technologies for scalable and distributed data processing. Established big data processing frameworks like Spark and MapReduce are indispensable to the bioinformatics practitioner. Keeping familiar abstractions and being able to reuse existing code with Intel SGX is important for its adoption. In the following, we will present how Intel SGX can be successfully deployed in the cloud, how it integrates with existing cloud technologies and briefly touch on open challenges.

4.1 TEEs in the cloud

Because of security concerns when moving computation to the cloud, it is only sensible to use TEEs to protect cloud applications. Although cloud providers go to great lengths to secure their infrastructure, TEEs offer an additional layer of security that the application developer controls exclusively. TEEs provide a secure environment for confidential computing in an otherwise untrusted public cloud. Even the proverbial disgruntled employee with physical access to the servers cannot extract customer data, for example, a genome that is currently analyzed, from within a TEE.

Nevertheless, the cloud provider retains full control over the cloud infrastructure. It is still responsible for allocating resources and running the associated services reliably. The rich ecosystem of integrated cloud services (database, key-value store, virtual network, coordination service, domain name service, message queue, load balancer, etc.) significantly reduces the operational overhead of deploying applications in the cloud. As genomic data processing routinely requires to analyze terabytes of data, it is a perfect fit for the cloud where computational resources scale conveniently with the size of the problem.

As Intel SGX-enabled processors were released in 2015, cloud providers have begun to integrate Intel SGX into their service offerings [27,28]. Due to SGX's novelty, we expect the integration into the existing cloud ecosystem to evolve over time. Now that different TEE implementations from multiple hardware vendors are available, cloud providers might be inclined to offer a unifying programming interface to abstract the details of the individual TEEs. We may also see software-only implementations of existing hardware TEEs that can be used as a fall back if the actual hardware might be unavailable. Cloud providers will continue to innovate in that space and it will be interesting to see how the integration of TEEs into cloud platforms evolves over time.

Even though SGX provides a strong trust anchor, the cloud user may still have to trust the provider to some degree with other aspects. For example, if it is important for the data to stay within certain geographic boundaries, the cloud user must trust the cloud provider to correctly implement "geo-fencing." SGX does not offer any mechanisms to enforce this. Today, the tier-one cloud providers typically have a global presence. Cloud users decide at which point-of-presence to instantiate a workload. As there is no way to verify the actual location of the virtual resources, the cloud user must trust the provider. Similarly, to truly leverage the power of the cloud, the cloud user might have to trust the cloud provider with other tasks too. Cloud architects are just beginning to reason about the security implications and trade-offs of TEEs in the cloud.

4.2 Developing with SGX in the cloud

Developing SGX applications to run in the cloud is similar to developing regular SGX applications in many aspects. The application developer splits the application in a trusted and untrusted part, defines the interfaces at the trust boundary, and follows established practices when developing security-sensitive SGX applications.

However, in other aspects, the cloud is also a very different environment from executing a single SGX application on a local machine.

For one, as multiple cloud applications from different cloud users share the same infrastructure, they must be isolated from each other. This is typically done using virtualization, possibly in combination with containers. Although SGX was designed with virtualization in mind, the hypervisor must support it. Patches exist for the popular open-source Xen and KVM hypervisors [29], and cloud providers are adding virtualized SGX support to their own hypervisors. Supporting SGX inside containers is less problematic as containers share the underlying host operating system. The container only needs access to the SGX device and be able to communicate with SGX daemons running on the host. If this is set up correctly, SGX applications can be launched inside containers without problems.

Because containers are a popular vehicle to package and deploy cloud applications, some initial work exists on automatically converting an existing container into an "SGX container" [30]. Static analysis of the container reveals the main binary and its runtime dependencies. With the help of Graphene-SGX [16], which allows to run existing binaries transparently on SGX, the main binary is hoisted into an SGX enclave. A new container image is constructed in which Graphene-SGX replaces the existing main binary as the entry point. Ultimately, if the resulting "SGX container" image is instantiated, the main binary will automatically run inside an SGX enclave. In this way, existing genomics workflows, including WDL-based execution pipelines [31], maybe converted with low effort to run on SGX.

Running an existing program on SGX using one of the mentioned solutions has another benefit. The solutions work by wrapping the unmodified program and mediating all the actions that are not permitted in an enclave. As SGX introduces a new processor mode, certain operations typically available in user mode are disallowed in enclave mode. The wrapping code can work around some of these restrictions. The wrapping code can also transparently encrypt any data leaving the enclave. As existing applications trust the environment, they typically output clear text data, either by writing it to a file or sending it over the network. The wrapping code can thus act as a "shield": it encrypts all data exiting the enclave and decrypts all incoming data. Besides transparent encryption, the wrapping code can enact a variety of security policies to retrofit security onto otherwise security-oblivious applications.

In summary, developing SGX applications for the cloud is similar to developing stand-alone SGX applications. However, virtualization, packaging, deployment and the desire to preserve existing workflows create additional challenges that must be overcome. The technical building blocks exist today, but it will take additional time for them to mature before cloud providers will want to integrate them into their infrastructure.

4.3 Deploying SGX applications

As already mentioned earlier, cloud applications tend to be a distributed collection of cooperating services rather than a single monolithic application. To help with the instantiation and orchestration of multiple cooperating services, cloud users write a "recipe" that states which services should be launched together. The recipe includes information on service dependencies (e.g., the database must be up before application server), their communication topology (e.g., the application server can talk to the database) and configuration information for each service. An orchestrator utility takes the recipe and starts all the listed services. The orchestrator also configures the (virtual) network to enable their communication and restarts services if they crash. Examples of these orchestrators include Docker Swarm [32], Kubernetes [33] and cloud-platform specific implementations like AWS CloudFormation Templates [34], Google Cloud Composer [35], and Microsoft Azure Automation [36].

For confidential computing in public clouds, it would be highly desirable to have a "secure" version of this orchestrator. The developer provides an existing recipe to the secure orchestrator. The secure orchestrator takes the existing applications and wraps them inside SGX enclaves; similar to the automatic conversion of containers into "SGX containers." The secure orchestrator uses the communication topology from the recipe and installs an equivalent security policy. The security policy is enforced at the enclave boundary where the wrapping code interacts with the untrusted outside world. For example, if the recipe states that the database can communicate with the application server, the policy would disable all network communication by default and only allow these two services to establish a network connection.

Besides restricting communication to known services listed in the recipe, the in-flight data must also be secured. To protect the in-flight data, the SGX-enabled application establishes an attested and authenticated communication channel between the participating endpoints. Attestation ensures that the enclave is, in fact, a genuine enclave running on an up-to-date SGX-enabled platform. The attestation serves to authenticate the enclave to interested parties. With RA-TLS [12] there already exists a solution to establish an attested secure channel between two SGX enclaves. RA-TLS uses the standard Transport Layer Security (TLS) protocol and integrates SGX remote attestation into it. RA-TLS solves the problem of secure interactive interenclave communication. RA-TLS can be used not only to send secrets, such as encryption keys, to the enclave but to protect communication between cooperating enclaves in general. For example, RA-TLS can secure the communication of a multistage WDL-based workflow. In this scenario, RA-TLS ensures that only the correct next stage (as measured through MRENCLAVE/MRSIGNER) is able to receive the input data from the previous stage.

Besides the ability to run custom applications in the cloud, providers also host an ever-increasing set of common processing frameworks. Map/Reduce, Function-as-a-Service (FaaS), stream processing (e.g., Spark), and data flow-based frameworks are sufficiently popular among cloud users for the provider to offer them as hosted solutions. Researchers have already started to integrate Intel SGX with these existing programming models to make confidential computing easily accessible to a wider class of users and application domains. For example, Microsoft Research ported Map/Reduce-style computations to Intel SGX and called the resulting system VC3 [37]. Similarly, Opaque [38] is an SGX-enabled Spark engine developed by the University of California, Berkeley. Opaque not only helps protect the confidentiality and integrity of the data but also obfuscates the communication patterns between components to prevent an attacker from learning sensitive information via side channels. The result is an oblivious database system where the query execution hides information about the distribution of the underlying data.

Unfortunately, current SGX-enabled data processing frameworks (e.g., both VC3 and Opaque) require to rewrite the code put inside the enclave in C/C++. Rewriting a large code from a high-level language like Python or Java into low-level C code may be impractical; in addition, genomics scientists may not have skills for such a venture. For wide adoption, it is imperative that SGX enclaves allow to run logic written in common data-scientist languages. In the future we expect more and more processing frameworks and language runtimes to integrate natively with trusted execution technologies. Cloud providers will then be able to offer secure, SGX-enabled versions of popular data processing frameworks.

4.4 Open questions on SGX in the cloud

Although most aspects of the current SGX ecosystem are easily transferable to the cloud, some specific concepts do not translate as well. We briefly highlight some facets of SGX that must still be aligned with the cloud's operational model. The topics mentioned here require further exploration and are areas of active research.

Sealing. Intel SGX helps protect the integrity and confidentiality of a program at runtime. For a complete solution, a program's inputs and outputs must also be protected at rest. To this end, SGX supports *data sealing*. Before storing data, the enclave encrypts it with a key only available on this particular platform. In this way, even if an attacker manages to exfiltrate the sealed data, the attacker will be unable to decrypt it, as instantiating the same enclave on a different platform will result in a different sealing key. However, sealing to the platform is of limited use in the cloud, as the infrastructure is virtualized. When an enclave is restarted, there is no guarantee that it will come up on the same physical server as before. Hence, protecting data from exfiltration and the ability to only be decrypted by authorized enclaves requires a different approach.

One option is to deploy a key management enclave. Application enclaves remotely attest to the key management enclave to access their decryption keys. Cloud service providers already offer key management services. They could either run their existing service unmodified on SGX, for example, Barbican [39] or develop a new key management service with a small TCB from scratch.

In general, for a distributed scalable SGX application like a key manager itself, instead of sealing the secrets to enclave, SGX sealing can be used to seal a shared master key. This key gets provisioned during instance creation. Each instance of the SGX application then seals its secrets using the shared master key.

Deep attestation. Sealing also involves the notion of enclave identity and deciding which enclave should be able to derive specific encryption keys. We already touched upon the topic of cloud applications often being distributed. Although an enclave's identity is defined through a cryptographic hash measurement of its code, a cloud application includes multiple SGX enclaves potentially with different identities/code measurements. Attestation is used by a relying party to decide on whether to trust an enclave. In the cloud, the relying party is deciding on whether to trust an ensemble of cooperating SGX enclaves. One possibility to achieve a deep attestation is to have the secure orchestrator issue a "cloud service attestation." This attestation is essentially a signed statement by the orchestrator that it faithfully instantiated and successfully attested each component. A client of the cloud service can inspect the "cloud service attestation" to determine for themselves the trustworthiness of the cloud service. In the genomics domain the aggregate service may be an entire processing pipeline where each pipeline step is a single SGX enclave. To gain trust in the entire pipeline, all stages must be attested. A potential solution must trade off the complexity and frequency of attestations as well as the desire to hide the cloud service's implementation details, among other things.

4.5 Putting it all together

Coming back to our example of aggregate association analysis, the application description given earlier was overly simplistic: instead of a single simple application, the analysis is more likely a workflow of chained transformation steps, for example, using the GATK [1], written in WDL [31], a language allowing to describe workflow graphs of steps running in containers. Additionally, analysts would not directly launch the application but use runtimes such as Cromwell [31] to dispatch and schedule the workflow in a cloud environment.

Fig. 8.12 illustrates how we could compute such a workflow in an untrusted cloud. Individual steps $f_x()$ will run as-is inside earlier-discussed SGX containers. A small shim in the container enforces that communication between steps is

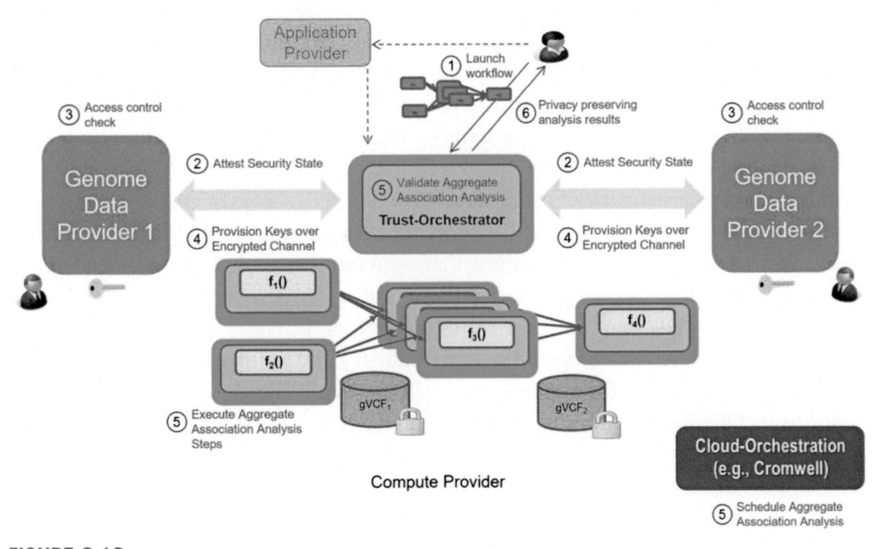

FIGURE 8.12

Example of application in Cloud.

properly secured using RA-TLS. The shim also handles the application-transparent encryption and decryption of data files. A separate new component, dubbed trust-orchestrator, proxies between all involved SGX enclaves and the different parties, whereby providing them with implicit deep attestation over the whole workflow. This simplifies the interaction and hides unnecessary details of the cloud topology from external parties. In conjunction with the shim in the container, the trust-orchestrator handles key management and, importantly, also enforces the integrity of the workflow graph while still leaving the scheduling decisions to the existing untrusted scheduler. This reduces the TCB size while also allowing the cloud operator complete control to optimize overall cloud resources.

5. Outlook

Intel SGX is available on current generation Intel client platforms as well as low end servers. There is lots of traction in the academic community with over 170 papers published on SGX over last 3 years. On the client side, the usages include password managers [40], secure web browsers [41] and SGX-enabled DRM [42]. Cloud computing is where a lot of the academic work is focused and usages include Genomics privacy-preserving analytics [24], distributed ledger [43], encrypted databases [44], middle boxes [45], and many more.

In addition to academic traction, there are a number of startups sprouting that are productizing SGX-based HSMs [46], key managers [39], network middle boxes [47] and more. There is an emerging trend called decentralized cloud computing [48] that relies on SGX TEE for running arbitrary computations on any computer participating in decentralized cloud computing. Microsoft Azure recently announced Azure cloud computing deploying Intel SGX as well as a software SDK called Open Enclave SDK [27]. Google Cloud Platform has announced Asylo [28], an open-source framework for confidential computing. Asylo allows development of applications at a higher-level abstraction and supports various backend like Intel SGX and AMD SEV. With various SDKs, cloud users have tools to build their secure applications, but still there is a strong desire to run unmodified legacy cloud workloads on SGX. There are a number of academic/production projects like SCONE [14], SGX-LKL [15], and Graphene-SGX [16], which provide support for running unmodified legacy applications inside an enclave. Going forward these environments will provide automatic protection of complex workloads using SGX, but users still have the flexibility to partitioning workloads that are sensitive to the size of TCB.

Even though there is clear traction in the industry and among the cloud providers and a clear path forward for applications such as secure outsourcing, there are still numerous technical challenges to the widespread adoption of TEEs. Adopting TEEs in multiparty settings, in particular for sensitive and highly regulated data such as health information, also raises additional nontechnical challenges: as any new technology, there is legal and regulatory uncertainty, for example, whether absolute or relative identifiability is required in the EU has a large impact on whether MPC, based on either hardware or crypto, is sufficient or not [49]. Additionally, a public acceptance that is based as much on technical aspects as it is on perception has yet to be tested.

Nevertheless, we consider TEEs and HW-MPC a very promising avenue to allow the pooling of sensitive data and unleashing of the vast potential of collaborative analysis.

References

[1] https://software.broadinstitute.org/gatk/.

[2] https://researcher.watson.ibm.com/researcher/view_group_subpage.php?id=2677.

[3] https://trustedcomputinggroup.org/work-groups/trusted-platform-module/.

[4] https://www.intel.com/content/www/us/en/architecture-and-technology/trusted-infrastructure-overview.html.

[5] https://keystone-enclave.org/.

[6] https://developer.arm.com/technologies/trustzone.

[7] https://globalplatform.org/.

[8] https://developer.amd.com/amd-secure-memory-encryption-sme-amd-secure-encrypted-virtualization-sev/.

[9] https://software.intel.com/en-us/sgx.

[10] Johnson, et al. Intel Software Guard Extensions: EPID Provisioning and Attestation Services. Intel Whitepaper; 2016. https://software.intel.com/en-us/blogs/2016/03/09/intel-sgx-epid-provisioning-and-attestation-services.

[11] Scarlata, et al. Supporting Third Party Attestation for Intel® Software Guard Extensions Data Center Attestation Primitives. Intel Whitepaper; 2018. https://software.intel.com/en-us/download/supporting-third-party-attestation-for-intel-sgx-data-center-attestation-primitives.

[12] https://github.com/cloud-security-research/sgx-ra-tls.

[13] https://developers.google.com/protocol-buffers/.

[14] Arnautov S, Bohdan Trach, Gregor F, Knauth T, Martin A, Priebe C, Lind J, Muthukumaran D, O'Keeffe D, Stillwell ML, Goltzsche D, Eyers D, Kapitza R, Peter P, Fetzer C. SCONE: secure Linux containers with Intel® SGX. OSDI; 2016.

[15] https://github.com/lsds/sgx-lkl.

[16] Tsai C-C, Vij M, Porter D. Graphene-SGX: a practical library OS for unmodified applications on SGX. USENIX ATC; 2017.

[17] Stephen C, Shacham H. Iago attacks: why the system call API is a bad untrusted RPC interface. ASPLOS; 2013.

[18] Wang W, Chen G, Pan X, Zhang Y, Wang XF, Bindschaedler V, Tang H, Gunter CA. Leaky cauldron on the dark land: understanding memory side-channel hazards in SGX. CCS; 2017.

[19] Van Bulck J, Frank P, Strackx R. Nemesis: studying microarchitectural timing leaks in rudimentary CPU interrupt logic. CCS; 2018.

[20] Jang Y, Lee J, Lee S, Kim T. SGX-Bomb: locking down the processor via Rowhammer Attack. SysTEX; 2017.

[21] Xu Y, Cui W, Marcus P. Controlled-channel attacks: deterministic side channels for untrusted operating systems. S&P; 2015.

[22] Szekeres L, Payer M, Tao W, Song D. SoK: eternal war in memory. SP. 2013.
[23] Ruan de Clercq, Verbauwhede I. A survey of hardware-based control flow integrity (CFI). arXiv; 2017.
[24] Chen F, et al. PRINCESS: privacy-protecting rare disease international network collaboration via encryption through software guard extensions. Bioinfornatics 2017;33(6): 871–8.
[25] https://github.com/intel/linux-sgx.
[26] https://www.hackread.com/amazon-suffers-security-breach/.
[27] https://azure.microsoft.com/en-us/blog/introducing-azure-confidential-computing/.
[28] https://cloudplatform.googleblog.com/2018/05/Introducing-Asylo-an-open-source-framework-for-confidential-computing.html.
[29] https://01.org/intel-software-guard-extensions/sgx-virtualization.
[30] https://github.com/cloud-security-research/graphene-sgx-secure-container.
[31] https://software.broadinstitute.org/wdl/.
[32] https://docs.docker.com/engine/swarm/.
[33] https://kubernetes.io/.
[34] https://aws.amazon.com/cloudformation/aws-cloudformation-templates/.
[35] https://cloud.google.com/composer/.
[36] https://azure.microsoft.com/en-us/services/automation/.
[37] Schuster, et al. VC3: trustworthy data analytics in the cloud using SGX. In: IEEE symposium on security and privacy; 2015.
[38] Zheng, et al. Opaque: a data analytics platform with strong security. NSDI; 2017.
[39] Chakrabarti S, Baker B, Vij M. Intel® SGX enabled key manager service with OpenStack Barbican. arXiv; 2017. https://arxiv.org/abs/1712.07694.
[40] https://pdfs.semanticscholar.org/ec40/833215b3d415c9525940690d0a94d2a178ca.pdf.
[41] https://software.intel.com/en-us/articles/hardening-authentication-tokens-in-browsers-using-intel-software-guard-extensions.
[42] https://software.intel.com/en-us/articles/using-innovative-instructions-to-create-trustworthy-software-solutions.
[43] Brandenburger M, Cachin C, Kapitza R, Sorniotti A. Blockchain and trusted computing: problems, pitfalls, and a solution for hyperledger fabric. ArXiv; 2018. https://arxiv.org/abs/1805.08541.
[44] Priebe C, Vaswani K, Costa M. EnclaveDB – a secure database using SGX. S&P. 2018.
[45] Poddar R, Chang L, Popa RA, Ratnasamy S. SafeBricks: shielding network functions in the cloud. NSDI; 2018.
[46] https://fortanix.com/.
[47] https://arxiv.org/pdf/1706.06261.pdf.
[48] https://enigma.co/enigma_full.pdf.
[49] Damiani E, editor. Evaluation and integration and final report on legal aspects of data protection; 2016. Deliverable D31.3, EU Project PRACTICE (Privacy-Preserving Computation in the Cloud).

Index

'*Note:* Page numbers followed by "t" indicate tables and "f" indicate figures.'

H

Printed in the United States
By Bookmasters